Biological Science

Maintenance of the organism
a Laboratory guide

Edited by C. F. Stoneman

Nuffield Advanced Science
Published for the Nuffield Foundation by Penguin Books

Penguin Books Ltd, Harmondsworth, Middlesex, England
Penguin Books Inc., 3300 Clipper Mill Road, Baltimore,
Md21211, U.S.A.
Penguin Books Ltd, Ringwood, Victoria, Australia

Filmset in Monophoto Ionic
by Oliver Burridge Filmsetting Ltd, Crawley and
made and printed in Great Britain
by H. L. Vickery Ltd, Hackbridge
Designed by Ivan and Robin Dodd
Illustrations designed and produced
by Penguin Education

Contents

Foreword   Sixth form courses in Britain have received more than their fair share of blessing and cursing in the last twenty years: blessing, because their demands, their compass, and their teachers are often of a standard which in other countries would be found in the first year of a longer university course than ours: cursing, because this same fact sets a heavy cloud of university expectation on their horizon (with awkward results for those who finish their education at the age of 18) and limits severely the number of subjects that can be studied in the sixth form.

So advanced work, suitable for students between the ages of 16 and 18, is at the centre of discussions on the curriculum. It need not, of course, be in a 'sixth form' at all, but in an educational institution other than a school. In any case, the emphasis on the requirements of those who will not go to a university or other institute of higher education is increasing, and will probably continue to do so; and the need is for courses which are satisfying and intellectually exciting in themselves – not for courses which are simply passports to further study.

Advanced Science Courses are therefore both an interesting and a difficult venture. Yet fresh work on Advanced science teaching was obviously needed if new approaches to the subject (with all the implications that these have for pupils' interest in learning science and adults' interest in teaching it) were not to fail in their effect. The Trustees of the Nuffield Foundation therefore agreed to support teams, on the same model as had been followed in their other science projects, to produce Advanced courses in Physical Science, in Physics, in Chemistry, and in Biological Science. It was realized that the task would be an immense one, partly because of the Universities' special interest in the approach and content of these courses, partly because the growing size of sixth forms underlined the point that Advanced work was not *solely* a preparation for a degree course, and partly because the blending of Physics and Chemistry in a single Advanced Physical Science course was bound to produce problems. Yet, in spite of these pressures, the emphasis here, as in the other Nuffield Science courses, is on learning rather than on being taught, on understanding rather than amassing information, on finding out rather than on being told: this emphasis is central to all worthwhile attempts at curriculum renewal.

If these Advanced courses meet with the success and appreciation which I believe they deserve, then the credit will belong to a large number of people, in the teams and the consultative committees, in schools and universities, in authorities and councils and associations and boards: once again it has been the Foundation's privilege to provide a point at which the imaginative and helpful efforts of many could come together.

Brian Young
Director of the Nuffield Foundation

General Editor of the *Laboratory guides*·
W. H. Dowdeswell

Editor of this volume:
C. F. Stoneman

Contributors:
C. F. Stoneman
W. H. Freeman
Margaret K. Sands

Preface The purpose of this *Laboratory Guide* is to provide a coherent treatment of a series of biological topics, based on laboratory investigations. The approach throughout is centred on enquiry as opposed to the mere verification of facts and through this work you will gain not only knowledge but also an understanding of the processes by which biological knowledge is acquired. Observation, the designing of investigations, working out ideas–hypotheses–on which to base experiments, analysing results, and drawing relevant conclusions will play a fundamental role in the work you do.

No matter how logically it may appear to hang together, a sequence of laboratory investigations inevitably leaves large and important gaps of knowledge uncovered. For this reason each investigation is preceded by a short *introductory section*, intended to give it relevance and relate it with what has gone before and what is to follow. Following this, detailed instructions are provided under the title *Procedure*. However, you will find that a good deal of latitude has been left for individual initiative and interpretation in designing and carrying out experiments. At the end of each investigation there is a short series of *Questions*. It is advisable to read these carefully before embarking upon the relevant investigation as, to some extent, they point the direction in which the enquiry should be conducted.

At the end of each chapter is a short and carefully selected *Bibliography*. This lists the titles of a few suitable books (it makes no pretence at being exhaustive) which will enable you to broaden your understanding of the topics you have investigated in the laboratory and see how they are related to others. At the same time the list indicates the range of topics that you should aim to cover in your reading.

In the short period of time available it is obviously impossible to span more than a fraction of the subject matter by practical work. So, just as in scientific research, we must acquire our knowledge both by laboratory studies and by using fully the findings and interpretations of others.

Each investigation in the *Laboratory Guide is numbered in the* lefthand margin, using a decimal system to allow easy cross reference.

Complementary to the series of *Laboratory Guides* is the *Study Guide*. This is also a book of investigations but it does not involve practical work and makes use of various types of second hand data instead. It serves also as a further source of theoretical information and is integrated with the work in the *Laboratory Guides*.

Synopsis

1 The examination of a model or a natural community provides information and poses problems about the distribution of animals and plants.

2 In an attempt to relate distribution with physical factors, one of these factors is measured. Any such relationship may prove much more complex than superficial examination would suggest.

3 The complexity of the relationship may be due, in part, to the effect that organisms have on their environment. Plants and insects can be suitable subjects for this kind of study.

4 The exchange of matter between organisms and their environment provides a material basis for complex interactions between communities and the habitats in which they exist.

Chapter 1
Interaction and exchange between organisms and their environment

Summary of practical work

section   *topic*

1.1   Distribution in a model or natural community

1.2   Concentration of oxygen in pond water

1.3   Interaction between organisms and their environment

1.4   Variation of carbon dioxide in air

1.5   The effect of gas changes on locusts' breathing

1.6   Human consumption of oxygen

Studying a model community

Biology is the study of living things and these are all around us. We ourselves are living organisms. In the neighbourhood of a school you can find a great number of different kinds of plants and animals.

A collection of organisms living together in a particular place is known as a *community*. Most communities include large numbers of individuals often belonging to different species and, at first sight, it is often difficult to decide where any useful study can begin. The situation is often made more difficult by the fact that the limits of a single community are seldom clear-cut but the community usually tends to merge with its neighbours to form a more or less continuous system.

Scientists in the past have made progress with large problems by first breaking them down into smaller ones. If we were to select an area such as a playing field or a lawn tennis court, it would be a mammoth task to obtain even an approximate idea of the distribution of plants and animals present on account of the number of species involved. So we would do well to begin with a small community. If you have a pond nearby, this may be suitable, but there is a lot to be said for starting with an artificial situation by setting up a model pond in a tank in the laboratory. This can contain water, mud, and a collection of organisms brought in from a local source.

A community like this usually presents us with a number of interesting problems. For example:
Why are there differing numbers of organisms?
Why are some populations* larger than others?
Do the populations depend on one another or are they independent?
Are the populations evenly mixed and distributed throughout the community or are they segregated? If they are segregated, what causes this?

Such a list can be extended almost indefinitely; we should be wise to limit our attention to the simpler questions first.

The most obvious problems are those relating to the distribution of species. The answers to such problems will not only tell us something about the pattern of distribution of populations within the community, but will also provide a starting point for a number of other questions concerning numbers, interaction between species, and the lives of individual organisms.

The problem of identification

If we are to study a community scientifically, we must have some adequate means of naming the various species present. At the outset it may be sufficient to use generalized terms such as 'snails', 'flatworms', 'water fleas', 'Algae', and 'pond weeds'. But at a later stage it will be necessary to carry out a proper identification, at least to genus level and sometimes to species. For this you will need to use keys.

* All the organisms belonging to one species, living in a community, constitute a *population*.

1.1 Distribution in a model or natural community

Even the most superficial examination of a freshwater *habitat* of plants and animals will show that distribution is uneven. (Habitat is a term meaning the natural home or habitation of a plant or animal and should not be confused with 'environment' – the sum total of the conditions in which an organism exists, including both living things and physico-chemical factors such as light and oxygen.) Flatworms, for example, may be found on the bottom or at the sides of a tank; water fleas are most likely to be swimming somewhere near the surface. A simple question can thus be asked: 'What is the distribution of one plant and one animal species?'

Procedure

1 From the tank or pond, select one plant and one animal species that appear to be unevenly distributed.
2 Remove a sample specimen of each.
3 Identify these using a key.
4 Make a record of the distribution of the two species. If you are using a tank, divide it for the purpose of recording into 'bottom', 'sides', 'water surface', and other regions. A diagram of a cross section of the tank, showing distributions, should be drawn.
5 Make a record of any variations in density of population. You can use words such as 'abundant', 'common', 'rare', 'absent', or you may find it preferable to assign arbitrary quantitative values to these variations.

Questions

a Did you find that the two species you selected each had a characteristic distribution, occurring in some places and being absent from others?
b Within the range of the two species, to what extent did their density vary?
c Could you correlate variations in distribution and abundance with the mode of life of the organisms concerned?

Variations in density of population

Having completed a preliminary survey of the distribution of two species, we may now be in a position to ask why they are present in some places and not in others. There are three kinds of possible answer:

1. The creatures in question may be unable to get into certain parts of the pond; they may not, for example, have the necessary means of locomotion.

2. They may be in one place rather than another by chance. If, for example, we see the entire population of six snails all on the sides of a tank, we should not assume that there is a reason for this any more than there is a reason for a tossed penny coming down heads.

3. Animals and plants may occupy their positions in the community because of the presence of food, oxygen, or some other vital material, or because a particular place affords shelter from predators or harmful light.

To say that an organism lives in a particular place because that is the best place for it, merely begs the question and does not help us to understand why. It is the third type of answer which leads to greater understanding and inevitably raises further questions of a profitable kind.

Many of the factors which might have a bearing on the distribution and life of organisms in a model pond are difficult to measure. A fairly easy one is oxygen concentration. Just as we need air to breathe, so fish need oxygen, dissolved in the water in which they live.

Though there may be no fishes in the model pond, it is reasonable to suppose that some of the animals present need oxygen, and consequently an uneven concentration of oxygen in the water may have a marked effect on their distribution.

1.2 Concentration of oxygen in pond water

There are several chemical methods of finding the concentration of oxygen in water; perhaps the best known is Winkler's method. A simpler technique is described here which requires less chemical knowledge and is sufficiently accurate and quick for our purposes. You should try it out first with a sample of water which has been well shaken for a few minutes to ensure that some air has dissolved in it. Once you have mastered the technique, you can test samples of pond water (see figure 1).

You must remember throughout that oxygen from the air may dissolve in your sample and that some already dissolved may escape from the water into the atmosphere. If either happens your determination will be misleading and so you should take steps to avoid such errors. Speed is essential.

A solution of iron(II) sulphate (ferrous sulphate) will decolorize certain dyes, one of which is phenosafranine. But if there is any oxygen present, the iron(II) sulphate reacts with it first, preferentially, and will not affect the dye until all the oxygen has been used up. Conditions for these reactions must be alkaline. This is the chemical basis of the following method.

Procedure

1 Obtain a sample of water (50 cm³) from one locality in the pond or tank, causing as little disturbance as possible. Use a large pipette or plastic syringe fitted with a disposable filling tube or length of plastic tubing. You may have to fill the syringe two or three times, depending on its size, but do not allow any air to enter it. Run this water into a small conical flask (capacity 100 cm³).

2 Add 5 cm³ sodium hydroxide (bench reagent).

3 Add 2 drops phenosafranine solution.

4 Attach a glass tube to a burette. This will give the burette a long jet whose opening will reach *below* the surface of the liquid in the conical flask. Fill the burette with a standard solution of iron(II) sulphate. Record the level of the meniscus in the burette and run a little of the solution into the flask below.

5 Swirl the mixture gently in the flask and if the colour still remains add more iron(II) sulphate solution from the burette. Continue this procedure until the colour of the dye just disappears.

6 Record the new meniscus level. By this time some of the colour may have returned but this does not mean that your estimate is incorrect. The previous remarks about dissolved oxygen should enable you to account for the return of the colour.

7 The above procedure is so adjusted that by using the appropriate concentration of iron(II) sulphate with a 50 cm³ sample, the volume recorded (second meniscus level minus

Figure 1
Diagram of the apparatus used for estimating the concentration of oxygen in water.

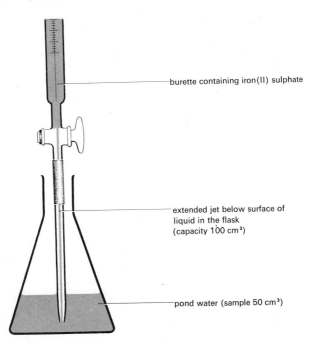

burette containing iron(II) sulphate

extended jet below surface of liquid in the flask (capacity 100 cm³)

pond water (sample 50 cm³)

first meniscus level) is almost exactly equivalent to the number of cm³ of oxygen per dm³ of water from which the sample is taken. Record this quantity. If you wish to convert this into percentage saturation, you may use the figures in table 1; you must first measure the temperature of the water at the point from which the sample was taken.

Temperature (degrees Centigrade)	Volume of oxygen (cm³) in water (1 dm³) at 0° C, 760 mmHg barometric pressure
8	8·13
9	7·95
10	7·77
11	7·60
12	7·44
13	7·28
14	7·12
15	6·96
16	6·82
17	6·68
18	6·54
19	6·40
20	6·28

Table 1
Oxygen contained in air-equilibrated* water at different temperatures.
*N.B. Under certain circumstances, in pond water, it is possible for there to be more oxygen dissolved, at a given temperature, than the amount stated above. Results of this kind, therefore, should not be regarded as incorrect.

8   Take samples from other places and depths and determine the concentrations of oxygen. Record these on a diagrammatic cross section of the tank as in the previous investigation.

Questions

a   What was the concentration of oxygen in the water sample that you tested?
b   How far did successive estimates of samples from a particular locality vary? How do you account for any variations?
c   Did the class results show any variation of dissolved oxygen in different localities? If so, can you suggest any hypotheses to explain them?

How to interpret the results
Concentrations of oxygen in tanks at different depths and places are unpredictable and often highly variable. In some tanks the concentration of oxygen is greatest near the surface; in others the reverse is true. One thing is certain: samples taken from the same place should give consistent results – your technique should be repeatable. Even so, it

may be argued that by taking a sample, you have altered the water in that place.

If there is a gradient of concentration or if there are local differences, see if these match differences in the distribution of animals or plants. Has your tank got abundant plant material floating on the surface of the water? Is one corner filled with pond weed? Is the bottom densely populated with water lice? In other words, do your determinations of the oxygen in any way correspond with the distribution of the organisms?

If so, you can propose that one of these conditions, distribution and the concentration of oxygen, causes the other. You have not proved any such connection but merely made a tentative hypothesis.

The next step would be to think of alternative hypotheses to account for the distribution of organisms and to try to devise experiments which demolish one or other of them. From one tank or pond alone, we could start dozens of lines of enquiry in this way and they would take us well outside the realm of distribution.

1.3 Interaction between organisms and their environment

If there is some connection between the distribution of certain animals in a tank and the local oxygen concentration, this raises the question, 'Does the oxygen concentration affect the distribution or do the animals affect the oxygen concentration?' If we consider our own environment instead of a pond, we know that if we remain in a small, sealed room, the air becomes changed and we suffer discomfort ultimately to the point of suffocation. We affect the environment and the environment affects us. The same is true for goldfish in a bowl or mice in a cage. In one way or another, animals change the environment around about them. Is this true for plants? Do the animals in a model pond ultimately suffocate if the tank is covered by a lid and sealed?

If we wish to determine the extent to which animals change their immediate environmental conditions we can do it fairly easily by using creatures which normally live in air. We can carry out a simple analysis of air by measuring the decrease in volume of a sample when we expose it first to potassium hydroxide solution and then to a solution of potassium pyrogallate. Potassium hydroxide absorbs carbon dioxide, pyrogallate absorbs oxygen. By keeping some organisms in a sealed container for a period of time and analysing the air at the beginning and end of this period, we should be able to get some idea of the amount of change the organisms bring about in the gaseous environment.

Procedure

1 Take some jars in which a number of locusts have been sealed for the past 24 hours. Obtain a sample of air by puncturing the lid of a container and quickly inserting the jet of a plastic syringe barrel (ideally, capacity 10 cm³). You should push the piston home fully before the operation and pull it back after the jet has been inserted.

2 Take an ordinary glass capillary tube with a bore of about 0·5 mm and fill it with water. Attach a short length of transparent plastic tubing to the sampling syringe and displace the normal air which it contains by the sampled air; then fix the transparent plastic tube to the capillary tube and introduce a bubble of air into it. When the bubble is about 10 cm long fit another plastic syringe (capacity 1 cm³) to the other end of the tube. Remove the larger syringe from the end of the capillary tube and place that end in water. By means of the smaller syringe, draw the bubble along the tube so that it is sealed by water at both ends (figure 2a). Alternatively, you can use a special capillary tube, bent in the shape of a letter J. One generally uses these with a screw device but they can also take a 1 cm³ syringe; either will do. With this, you can obtain a gas sample over water, as shown in figure 2b.

3 Leave the tube containing the bubble in water in a suitable container for at least five minutes. Note the temperature of the water. You must do this *before* you measure the volume because the volume of a gas sample varies according to the temperature.

4 Measure the length of the bubble to the nearest millimetre. If you wish to remove the capillary from the water container to do this measurement, speed is essential. Do not hold the glass tube in your hand or the rise in temperature will cause the gas bubble to expand. Record the length of the bubble.

5 By means of the 1 cm³ syringe (or screw device, if fitted), move the bubble towards the open end of the tube until it is about 1 cm from it (figure 2c). Put the open end into a solution of potassium hydroxide and move the bubble back to the opposite end of the tube. Still holding the open end in the hydroxide solution, shunt the bubble slowly back and forth along the length of the tube, at least five times.

6 Put the capillary tube back in the sink or water bath for five minutes and then measure the length of the bubble. Record this length.

7 Perform stages (5) and (6) again but use a solution of potassium pyrogallate instead of potassium hydroxide. Note that the pyrogallate is kept under a thin layer of paraffin oil. Plunge the open end of capillary through this quickly to prevent oil from entering the tube. Record the final length of the bubble.

8   Wash out the capillary tube with water, dilute acid, and water again to remove all traces of hydroxide and pyrogallate.

9   Using the same method, analyse air samples taken from a container of germinating seeds or seedlings and from an empty container. Air without organisms in it serves as a 'control'. It is useful to compare an analysis of this with analyses of the other samples. You can obtain some idea of the way we affect our environment by breathing into an empty (flat) polythene bag and taking a sample of this air. Though it is not strictly comparable with air from a sealed container of organisms, a sample of breath provides us with a simple means of increasing the scope of this investigation.

How to interpret the results

Suppose the original length of the bubble, in one of the analyses, was 100 mm, and when it was exposed to potassium hydroxide solution it shrank to 96 mm. In that case, the amount of carbon dioxide originally present was

$$\frac{(100-96)}{100} \times 100 = 4 \text{ per cent}$$

If, when this bubble was exposed to potassium pyrogallate solution, it shrank still further to 80 mm, then the content of oxygen of the original sample was

$$\frac{(96-80)}{100} \times 100 = 16 \text{ per cent}$$

The following analysis of London air is provided for comparison with your results from control samples. There is no reason why these results should be exactly the same as London air. Your control air may have been sampled, for instance, from a crowded laboratory.

Gas	Percentage volume
Oxygen	20·80
Carbon dioxide	0·03
Nitrogen	77·32
Argon	0·93
Water vapour (this varies with the weather)	0·92
Inert gases other than argon	0·005

The amount of carbon dioxide (0·03 per cent) is, of course, too small to be measured by this capillary tube method. A way of overcoming this limitation is mentioned in the next section.

As in the case of the previous titration method, the analysis of identical samples should produce identical results. In

Figure 2
Using a capillary tube for gas analysis.

a Introducing a gas sample into a straight capillary tube

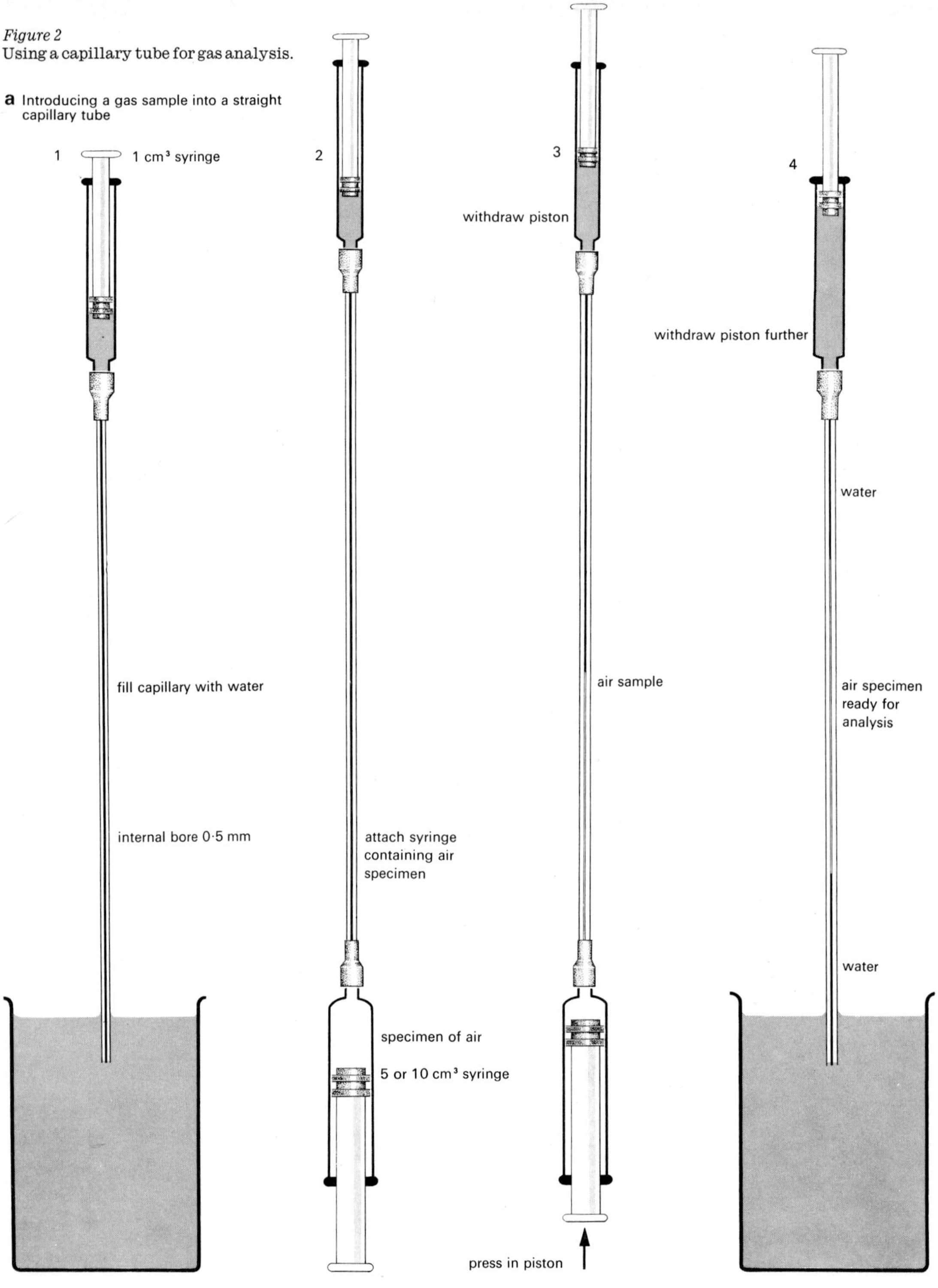

1   1 cm³ syringe

2

3   withdraw piston

4   withdraw piston further

water

fill capillary with water

air sample

air specimen ready for analysis

internal bore 0·5 mm

attach syringe containing air specimen

specimen of air

5 or 10 cm³ syringe

water

press in piston

b Introducing a sample of gas from an inverted syringe into a J-tube

c Preparing the sample of gas for analysis

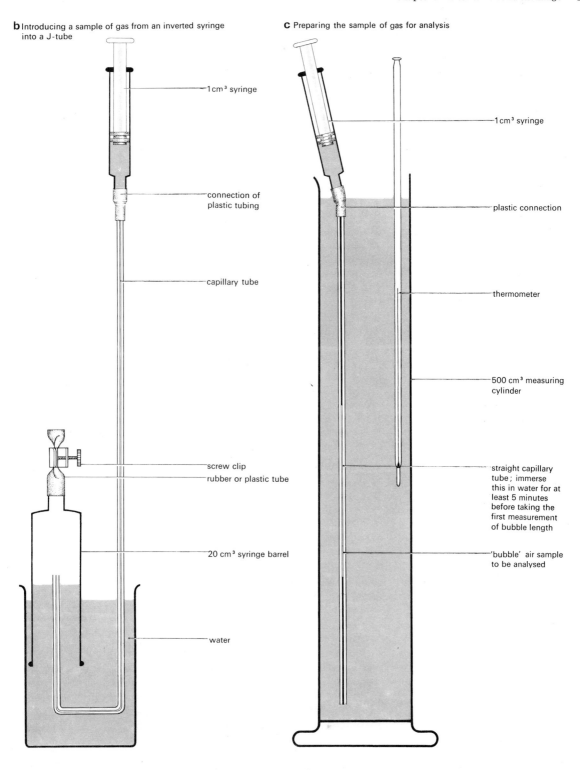

1 cm³ syringe

connection of plastic tubing

capillary tube

screw clip

rubber or plastic tube

20 cm³ syringe barrel

water

1 cm³ syringe

plastic connection

thermometer

500 cm³ measuring cylinder

straight capillary tube; immerse this in water for at least 5 minutes before taking the first measurement of bubble length

'bubble' air sample to be analysed

practice, even with experienced operators, this seldom happens since all measurements are subject to errors. The importance of estimating error in experiments is covered in all elementary books on statistics (see Bibliography). Reference to one of these will enable you to calculate the *standard deviation* of your class results. From twenty results obtained with the capillary tube method, one operator found that his standard error for the percentage of oxygen was ± 0.45 per cent. If your results have a much larger standard error you should examine your technique to see if it is reliable.

Questions

a   Compare the figures for oxygen and carbon dioxide obtained from air associated with locusts, germinating seeds, and human breath and from unaltered, laboratory air (control). Which set of results differs most from that obtained for laboratory air?

b   Compared with normal, laboratory air, do the other samples show: an increase in oxygen and an increase in carbon dioxide, or an increase in oxygen with no change in the amount of carbon dioxide, or a decrease in oxygen with an increase in carbon dioxide, or a decrease in oxygen with no change in the amount of carbon dioxide? (As already mentioned, no decrease in carbon dioxide, compared with 0.03 per cent, can be measured by this method.)

c   Was the answer to question (b) expected? If so, for what reason or reasons?

d   What further measurements would you have to make in order to find out the amount of change in the composition of air produced per unit of living organisms by locusts, germinating seeds, and human beings?

1.4 Variation of carbon dioxide in air

It is a serious drawback that we cannot measure the percentage of carbon dioxide in ordinary air by the capillary method. We can overcome it by using the acid property of carbon dioxide and the effect of acid on indicator dyes. Other gases in normal air do not affect indicators. However, in industrial areas, the air often contains some acid sulphur dioxide, and this will distort the results.

An M/1000 solution of sodium bicarbonate (containing 0.084 g 1^{-1}) is in equilibrium with air containing 0.03 per cent carbon dioxide; that is, the tendencies of carbon dioxide to dissolve into, and escape from, the solution are equal. If the dyes, cresol red and thymol blue, are mixed in this bicarbonate solution, the mixture is orange/red. When it is exposed to ordinary air, the colour remains the same. If there is more carbon dioxide than normal in the air above such a solution, some of it dissolves, the pH decreases, and the colour changes to yellow. If the carbon dioxide content is less than normal, then the pH of the solution rises and the indicators change to red and then red/blue.

Any acidic or alkaline substance can cause a colour change; you should bear this in mind when you use bicarbonate/indicator solution as a test for carbon dioxide.

Procedure

1 Pour 10 cm³ bicarbonate/indicator into each of four large test-tubes.
2 Cut discs of perforated zinc which are a little wider than the diameter of the test-tubes and fit them as shown in figure 3.
3 Weigh a locust, place it in one tube, and fit a rubber bung.
4 Put an equal weight of germinating seeds or seedlings into a second tube and insert a bung.
5 Exhale through a tube which nearly reaches the surface of the bicarbonate/indicator in the third test-tube. Withdraw the tube quickly and insert a bung.
6 Put a bung in the fourth test-tube but do not alter the air inside. Let it serve as a control for the purpose of comparison.
7 Check that the colour of the bicarbonate/indicator is the same in all the test-tubes when you first set up the experiment, and then compare them after ninety minutes and again after twenty-four hours.

Questions

Figure 3
Tubes for testing living things for the production of carbon dioxide.

a Assume that the changes in colour are due to carbon dioxide alone. Which of the test-tubes contains more carbon dioxide after one and a half and twenty-four hours, respectively? Suggest a reason for this.

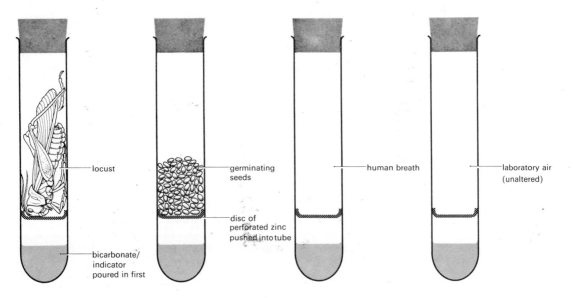

locust

germinating seeds

human breath

laboratory air (unaltered)

disc of perforated zinc pushed into tube

bicarbonate/ indicator poured in first

b   Has the locust caused the same change in colour as the germinating seeds?
c   Is it reasonable to expect the same change in colour when equal amounts of bicarbonate/indicator are exposed to equal weights of living materials for equal periods?
d   What factors outside the test-tubes might influence the rate of carbon dioxide production by the organisms?
e   How could you modify this technique so that it is more quantitative?

1.5 The effect of gas changes on locusts' breathing

Living organisms change the conditions of their environment; a change in the composition of the atmosphere is merely one example. Do such changes affect the organisms? In the model community, the concentration of oxygen in the water may have affected the distribution of organisms, but this is not quite the same thing as changes in the organisms themselves—although the two may be related.

We alter the atmosphere around us by breathing and so, for example, do locusts. If we look closely at a resting locust we can observe rhythmic movements of the abdomen. By these movements, it takes air into and out of the body through holes called *spiracles*.

If we assume that locusts obtain oxygen from the air by breathing then it would seem reasonable to suppose they would breathe faster as the concentration gets less. This is a simple hypothesis which you can test quite easily. You should have a resting locust in a transparent container and supplies of oxygen and carbon dioxide.

Procedure

1   Put an adult locust into a plastic syringe barrel (capacity 20 cm³). Insert the piston and push this in gently until the locust cannot move. Do not squash the animal!
2   Count the number of pumping movements of the abdomen which occur in thirty seconds. Record the results of three such counts in a total of one and a half minutes.
3   Attach a tube to the jet where the hypodermic needle is normally fixed, remove the piston without letting the locust escape, and breathe gently through the tube for ten to fifteen seconds. Push in the piston as before and count breathing movements for three successive periods of thirty seconds each. (The previous experiment will have indicated one of the components of breathed-out air.)
4   Remove the tube and move the piston to and fro about ten times so as to remove the breath and replace it with fresh air. Count the breathing movements of the locust for three periods of thirty seconds.

5   Remove the piston without letting the locust escape and re-place it with a plug of cottonwool. Gently pass pure oxygen, from a cylinder, into the syringe barrel. (Five to ten seconds should be sufficient.) Replace the piston and count the breathing movements as before.

6   Flush the syringe barrel with pure oxygen as before (5) and then add 5 cm³ carbon dioxide. Various ways of doing this are illustrated in figure 4. After recording the breathing movements for three successive periods of thirty seconds, add more carbon dioxide if there is enough room in the barrel. Count the breathing movements as before.

Questions

a   What effect, if any, does exhaled air have on the frequency of the breathing movements of a locust?

b   Does this result support the hypothesis that the movements become faster as the concentration of oxygen becomes less?

c   What conditions around the locust have been altered because you breathed into the syringe barrel? Do you have adequate means of knowing all the changed conditions?

d   How can you modify stage (3) of the procedure in order to reduce the number of variable factors associated with exhaled air?

e   Do the results of stage (6) of the procedure support the hypothesis or do they run counter to those of stage (3)?

f   Formulate a hypothesis about the rate of breathing and the composition of the atmosphere, from your results.

1.6 Human consumption of oxygen

We suffer discomfort and distress if we remain in a small un-ventilated space. In this way we are aware that we change our gaseous environment, but discomfort is rather a vague and subjective thing to measure. We could investigate the rate of a person's breathing in this situation by a technique similar to the one used on locusts but this would mean build-ing a large container and putting someone in it. This is a totally undesirable and dangerous procedure.

The established equipment for such investigations is called a spirometer and there are now several versions available.

Essentially, the apparatus consists of a box or bell (with a capacity of six or more dm³) suspended freely over water and counterbalanced so that gas passed in or drawn out makes the box rise and fall. (See figure 5.)

A pen is attached to the box and writes on a slowly rotating drum, thus making a permanent record of all the movements of the box.

16

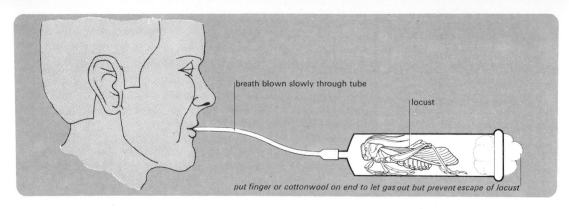

breath blown slowly through tube

locust

put finger or cottonwool on end to let gas out but prevent escape of locust

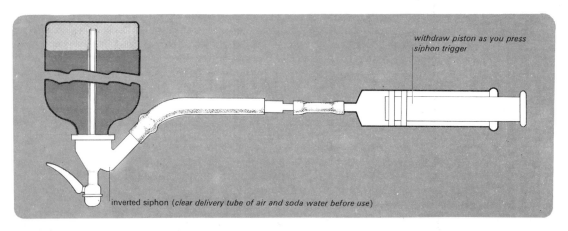

withdraw piston as you press siphon trigger

inverted siphon (*clear delivery tube of air and soda water before use*)

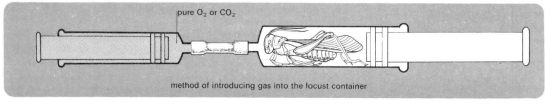

pure O_2 or CO_2

method of introducing gas into the locust container

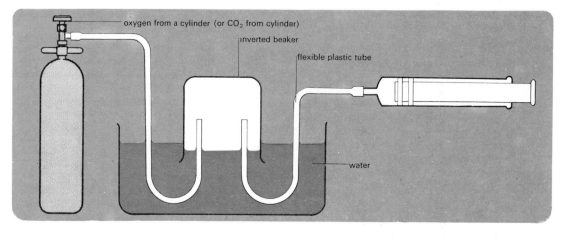

oxygen from a cylinder (or CO_2 from cylinder)

inverted beaker

flexible plastic tube

water

Figure 4
Methods of changing the gaseous
environment of a locust.

If a person (the 'subject') has his nose and mouth connected to such a spirometer containing pure oxygen, the box will go up and down, recording his rate of breathing. Though the movements of the box may change in frequency (the rate of up and down movements), the amplitude (the distance from highest to lowest position) will stay more or less the same. The reason for this is that as oxygen is being removed by the subject it is being replaced by an equivalent volume of carbon dioxide. This procedure is not particularly instructive (nor is it pleasant for the subject) for although it enables us to observe the frequency of breathing, it is not possible to measure the composition of the gas in the spirometer.

We can carry out a useful investigation by removing carbon dioxide from the system. In this case, the box will rise and fall with each breath and also subside gradually as the volume of oxygen decreases. This enables us to measure the subject's consumption of oxygen. We can, in this way, find out the rate at which human beings change the gaseous environment around them; the analysis of air samples taken in a capillary tube (1.3) merely shows that the composition of air does change during the act of breathing.

N.B.
Pupils should use spirometers only under the direct supervision of a teacher. Never *experiment on a human subject when you are not being supervised.*

Procedure

Figure 5
Diagram of a box-type spirometer.

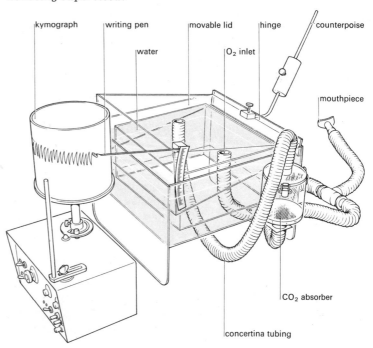

kymograph · writing pen · movable lid · hinge · counterpoise
water · O_2 inlet
mouthpiece
CO_2 absorber
concertina tubing

1   Consult the instructions provided by the manufacturer of the spirometer. Check that all the parts are present, assemble them, and make sure that you have an adequate stock of a chemical for removing carbon dioxide (e.g. 'carbosorb') and of *medical grade* oxygen. This is supplied in special cylinders which do not fit the ordinary reducing valves in general use in laboratories.

2   See that the pen writes on the drum of the kymograph all the time the drum is turning and for the whole of one turn. The rate the drum turns at should be less than one revolution per ten minutes so that you can make two five minute recordings on one sheet of paper.

3   Check for leaks by filling the spirometer with air, closing all taps, and loading the box with a weight (200 g). If the box does not move during five minutes, there are no serious leaks.

4   Calibrate the spirometer by letting the pen write on the kymograph drum when the box has sunk as low as possible; then let in a specific volume of air, e.g. 250 cm^3, and let the pen write on the drum for one revolution. Do this again, letting in the same volume of air each time until the box has risen as high as possible; then you will have calibrated the spirometer for the whole of its potential volume. Figure 6 shows a simple way of introducing a standard volume of air. When the pen is at its highest point, let the drum rotate and move the pen slightly once every ten seconds. This provides a time scale on the paper.

5   When you have finished calibrating and making all these checks, flush the spirometer with oxygen four times and leave it full, with the taps closed.

6   Find a subject. So that you can compare the results later with those in the preceding investigation, he (or she) should have rested for at least half an hour before the investigation and, ideally, he should not have eaten a meal during the previous two hours. When he is sitting comfortably, with the mouthpiece in position, begin recording at the *end of an outward breath* and continue for five minutes.

7   For additional recordings of interest, the subject should perform some vigorous exercise such as running up a flight of stairs. Recharge the apparatus with oxygen and see that the subject puts the mouthpiece in position again as soon as the exercise has ended and he is sitting down. Again, begin recording at the end of an outward breath.

How to interpret the results

From the vertical movement of the pen (from one peak to the next trough on the trace), you can determine the depth of the volume of breathing.

The rate of frequency of breathing may be determined by reference to the time scale.

Figure 6
A method of injecting a standard
volume of air.

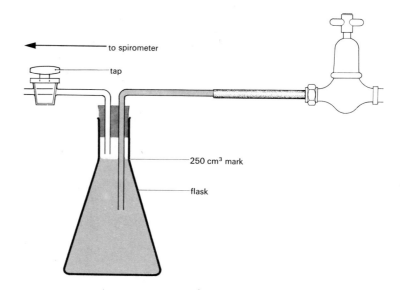

To find the amount of oxygen consumed, compare the lowest
point of the trace at the start of the five minute period with
the lowest point at the end of this period. Refer this vertical
distance to the adjacent calibration lines.

If a line drawn through all the peaks of the trace, in any one
period, is straight, this shows that the subject consumed
oxygen at a constant rate.

Questions

a   Would you expect the rate of consumption of oxygen, after
exercise, to differ from the rate in a resting subject? Give
reasons for your answer.

b   Determine how much oxygen the resting subject consumed
in five minutes and calculate how much he would consume,
at the same rate, in twenty-four hours. Compare this with
the amount of oxygen the locusts consumed in the earlier
investigation (1.3).
Do you need additional information in order to make this
comparison? If so, state what you require.

c   Calculate the rate at which a subject consumes oxygen after
vigorous exercise. Compare this result with the rate for a
resting subject. Is the amount consumed after exercise
greater than, the same as, or less than in a resting subject?

d   Are the results of the two measurements (stages 6 and 7)
strictly comparable? Are the conditions the same for both
measurements? If not, state the factors which vary. What
steps could be taken to overcome this variability?

Bibliography

Clegg, J. (1965) *The freshwater life of the British Isles*. Warne. (A good introduction to freshwater life with chapters on the roles of animals and plants in communities.)

Eggleston, J. F. (1970) Nuffield Advanced Biological Science Topic Review *Thinking quantitatively, I, Models and hypotheses, II, Statistics and experimental design*. Penguin. (Calculation of standard error.)

Garnett, W. J. (1965) *Freshwater microscopy*. Constable. (The use of a microscope applied to freshwater organisms.)

Harper, J. L., ed. Gray, J. H. (1970) Nuffield Advanced Biological Science Topic Review *Interactions*. Penguin. (The reaction of a plant to its neighbours.)

Macan, T. T. and Worthington, E. B. (1951) *Life in lakes and rivers*. Collins. (Chapters on photosynthesis and respiration and economic aspects of aquatic organisms.)

Sands, M. K. (1970) Nuffield Advanced Biological Science Topic Review *Control of breathing*. Penguin. (Effect of gas changes on rate of breathing.)

Synopsis

1 The fact that plants, insects, and human beings can change the composition of the air around them, prompts a consideration of the path along which gases are exchanged.

2 Leaves and locusts provide suitable specimens for studying the surfaces and internal structures where gas exchange takes place in a living thing. Use is made of material specially prepared for studying through a microscope.

3 Dissection of animals is an essential technique for obtaining information about anatomy which is complementary to physiological knowledge. It is not practicable to dissect a human being in schools, so mice, rats, or other small animals are used instead.

4 The capacity of human lungs can be found, without dissection, and the information obtained throws some light on the gas exchange system.

Chapter 2

Gas exchange systems

Summary of practical work

Plants and animals change the composition of the air around them. To do this their bodies must be open to the passage of gases, but this is complicated because most land organisms have some kind of skin (epidermis) which is more or less impermeable to water and also proof against gases. The skins or outer surfaces of plants and insects have an additional waterproof layer (cuticle) which makes them even more impermeable.

2.1 Gas exchange in leaves

It is easy to demonstrate how air passes through the epidermis of a leaf. Leaves of land plants contain air, as we can deduce from studying leaf sections. The air within a leaf can be forced out by heating the leaf and causing the air to expand.

Procedure

1 Heat a beaker of water until it has boiled for about one minute.
2 Remove the source of heat and, as soon as the last bubbles have reached the surface, plunge a leaf of *Impatiens balsamina* (busy Lizzy) under the surface.
3 Observe bubbles leaving the surface of the leaf.

Questions

a Do the bubbles come from the upper surface only,
b from the lower surface only, or
c from both surfaces? (The terms 'upper' and 'lower' refer to the natural position of the leaf on the plant.)
When you have made an observation about how gas escapes, try to see if the anatomy or structure of the leaf has some bearing on this.

Procedure

1 Pick another leaf of *Impatiens* and examine both the upper and lower surfaces under a microscope with a low power objective ($\times 100$).
2 Tear the leaf in two so that part of the thin, transparent skin (epidermis) of the underside is exposed. Mount a piece of this in a drop of water on a slide and cover with a cover-glass. Examine under low power and high power ($\times 100$, $\times 400$). (We shall usually refer to low and high power as L.P. and H.P. from now on.)
3 You will find it difficult to tear away a similar fragment of the upper epidermis. To obtain information about this it is best to make an impression (as in stage 4 below) or a replica (stage 5) of the surface.
4 Put a drop of clear nail varnish or, better, a drop of polystyrene cement on the upper surface of another leaf while it is still on the plant. Spread the varnish or cement thinly with a mounted needle. Allow it to dry (five to ten minutes). Cut off the leaf and immerse it in water in a Petri dish. Peel off the varnish or cement with a forceps, turn it over, and place it in a drop of water on a slide. Blot it firmly with filter paper and examine it under a microscope ($\times 100$ and $\times 400$). A cover-slip is not required.
5 An alternative worth trying is as follows. Pour a little ($1\,cm^3$) Silcoset 105 (a rubber latex) onto a slide or watch glass. Add one drop of Curing Agent 'D' and stir well with a matchstick. Quickly apply the mixture to the upper surface of a leaf. There is no need to spread it thinly; it will harden rapidly. Peel off this pad of rubber; put it on a flat surface with its epidermal impression uppermost and paint on a thin film of clear nail varnish. Allow this to dry (five to ten minutes). Peel the varnish off and mount it in a drop of water as in stage (4) above.
6 Insert a micrometer graticule into the microscope eyepiece and calibrate this with a stage micrometer slide using L.P. and H.P.

Questions

a Can you find holes (stomata) in the epidermis (see 1) or epidermal strip (see 2)?

b Are stomata to be found in the upper epidermis only, the lower epidermis only, or both?

c Do these answers correspond to those to the previous set of questions?

d How do the epidermis cells of *Impatiens* differ from those shown in figure 7, which is taken from the leaf of another kind of plant? Make a simple outline drawing to show about the same number of cells as in figure 7 but taken from your preparations.

e The stoma itself may be easily confused with the larger depression which surrounds it. Measure the length and greatest width of an open stoma in your preparation.

Figure 7
Cells of a leaf epidermis.
Photo, Rank Audio Visual.

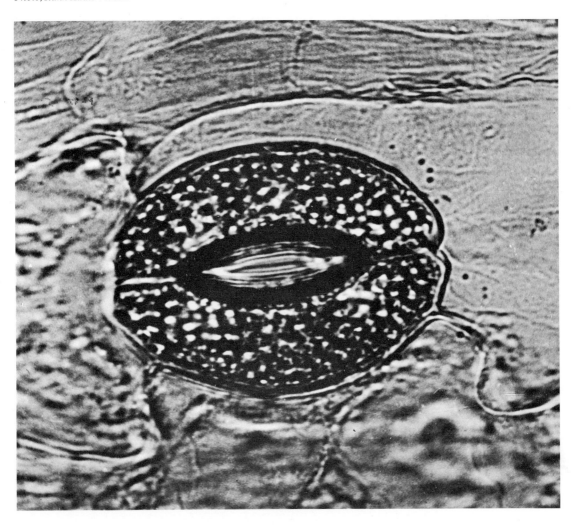

In order to complete the investigation and follow the passage of gases through the epidermis, you need to see something of the internal anatomy of a leaf.

Cross sections are difficult to cut by hand, and the following instructions apply to sections which have been cut by a microtome, stained, and mounted. If you cannot get slides of *Impatiens*, sections of privet leaf (*Ligustrum*) will do instead.

Procedure

1 Examine a cross section of a leaf under L.P. and H.P. Distinguish between the upper and lower surfaces by looking at the midrib which appears at the centre of the flattened 'V' of the leaf section.
2 Note the variety of cell shapes present and try to find a stoma in the lower epidermis.
3 With the eyepiece micrometer graticule measure the thickness of the leaf and also try to get some idea of the distance from a stoma to the inner wall of the upper epidermal cell above it. See distance d in figure 8.

Questions

a How many different kinds of cell, judging by their shape, can you find in the leaf section? Describe their shapes and positions briefly.
b What is the distance d (figure 8)?

Figure 8
Diagram of a vertical leaf section.

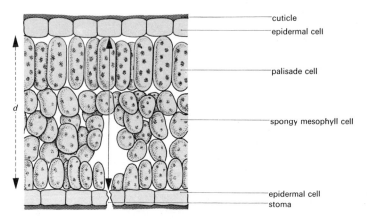

cuticle
epidermal cell

palisade cell

spongy mesophyll cell

epidermal cell
stoma

2.2 The breathing system of a locust

Leaves are covered with a waterproof layer, perforated by many small holes through which gases can pass. What about locusts? In the previous chapter it was found that locusts not only changed the atmosphere round them but also performed rhythmic movements which altered with changes in the composition of the atmosphere. The outer covering of a locust's body is shiny; it does not absorb drops of water and it would be unwise, morever, to assume that the whole surface is permeable to gases. If the locust performs breathing movements and changes the atmosphere correspondingly, then there must be holes or pores through which gases can pass.

Locusts have no obvious nostrils like mammals but gases may pass through their mouths. When you examined locusts in the previous chapter you may have discovered other apertures in the body surface. During the next step in this investigation we must make a more careful study of such apertures and dissect the animal to see what lies beneath the surface.

The aim of dissection is to find out what organs exist within a dead animal and to discover the anatomical relationships between them. We do this by opening the animal where it will do least damage and carefully displacing the internal organs. The dissector should not remove tissues unless they obscure more important parts or are required for further examination, for example, by microscopy. Dissection can provide information complementary to that which we obtain from experiments on living animals. Either kind of investigation may lead us to form hypotheses, but in order to give a complete account of a process, such as breathing, we need to understand both structure and function.

Procedure

1   Put a living, adult locust in a specimen tube so that its movements are restricted and examine it under a binocular microscope; look for likely apertures through which gas could pass.

2   Fix a dead, adult locust to the stage of a monocular microscope and examine apertures found in (1) more closely under L.P. ($\times 50$ or $\times 100$). An oblique source of light will help in this difficult investigation. It is not easy to distinguish between a true aperture and a pit or depression.
Record the positions of apertures by a simple outline sketch, for future reference.

3   Take a freshly killed adult locust and place it in its normal resting position in the centre of a shallow dish which has a layer of hard wax at the bottom. Fix it, by means of pins, through the legs and thorax. Cut off the wings if they get in the way.

4   With a forceps pull back the tip of the abdomen, stretching it slightly, and insert a pin to hold it back. With fine, pointed scissors, pierce the body covering (exoskeleton) between two segments at about the position shown in figure 9. Cut down between the segments on one side and then continue the incision forward to the head along the dotted line shown. As you cut, let the lower blade of your scissors pull the locust up slightly. In this way you will be less likely to damage internal organs.

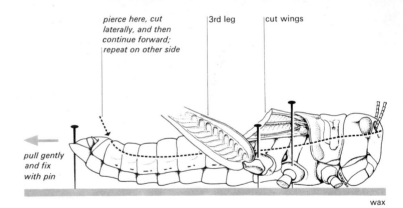

pierce here, cut
laterally, and then
continue forward;
repeat on other side

3rd leg

cut wings

pull gently
and fix
with pin

wax

Figure 9
A locust prepared for dissection –
side view.

5   Make a similar cut on the other side and carefully remove
     the strip of exoskeleton from the animal so that you can see
     the internal organs. (Use a blunt seeker to separate the
     strip from other tissues.) Preserve the strip in a specimen
     tube of alcohol for future use.

6   Explore the cavity and its contents with a seeker; identify
     the most prominent organ, the alimentary canal or gut
     (figure 10). You may separate it partly from the rest of the
     body but do not try to pull it away completely; it will appear
     to be 'tied' down by many small threads.

7   The main point of this dissection is to trace the path of gases
     into and out of the body. We are looking in particular at
     structures filled with air. Here we can make use of a pheno-
     menon which appears when an empty test-tube is placed in
     a beaker of water. It looks silver, having a mirror-like sur-
     face. Pour water into the dish until it covers the dissected
     body.

8   Examine the many silver threads and the associated organs
     with the aid of a hand lens and a binocular microscope. Pull
     the mid gut upwards gently and look at the air-filled tubes
     or *tracheae* from the side (figure 11).

Figure 10
A dissected locust showing the
alimentary canal.
After figure 27, Thomas, J. G. (1963)
Dissection of the locust, *Witherby*.

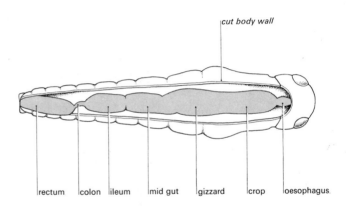

cut body wall

rectum   colon   ileum   mid gut   gizzard   crop   oesophagus

Questions

a Describe the different silver structures in the locust's body.

b List the internal parts of the body with tracheae leading to them. Examine, for example, the muscles and the ovaries or testes. (The ovaries are prominent yellow bodies; the testes are white, finger-like structures in the abdomen.)

c Describe the general arrangement of tracheae and their relationship with the apertures or *spiracles* in each segment. This is a difficult task; to answer the question completely demands skilled dissection and much time. The relationship between tracheae and spiracles is fairly straightforward. Record your findings in diagrammatic form, that is, a plan which shows the relative positions but does not necessarily make the organs look like the real ones.

d Is there any anatomical evidence to support the idea that locusts breathe through their mouths?

Figure 11
Side view of a dissected locust with the gut displaced.

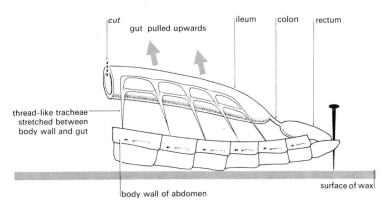

2.3 Fine structure of the breathing system of a locust

Dissecting the locusts has shown us that there are tubes connecting spiracles with internal organs. We are still not sure how air passes along them or what happens when the tracheae disappear into organs.

Procedure

1 With scissors, remove a small portion (5–10 mm) of the mid or hind gut together with some tracheae. (Figure 11.)

2 Cut the portion of gut longitudinally, open it out flat, wash away all traces of its contents, and mount the specimen in water on a slide beneath a cover-slip.

3 Observe it under L.P. and H.P. The tracheae, being filled with air, will stand out clearly. You can also tell them by the characteristic fine, transverse rings of material round them. Follow the course of a large trachea until it divides and follow this through successive divisions until you come to the end of one of the finest branches. Record these observations by precise drawings, accurate copies of the real structures.

4   Remove the cover-slip and put a drop of methylene blue (dissolved in water) on the piece of gut wall. This, or any other general stain, will help to make more features of the gut wall visible under the microscope.

5   While you are waiting for the stain to take effect, obtain a small leaf of *Impatiens* (busy Lizzy), cut out the midrib, tease it into shreds with needles, mount in a drop of water, and squash it firmly under a cover-slip. Examine under L.P. and see if you can find any structures which have a ringed appearance similar to that of tracheae.

6   Return to the preparation of gut wall, wash away the excess stain, and mount in water. Look for nuclei and other signs of a cell structure surrounding the tracheae (figure 12). Measure the diameters of the largest tubes.

Questions

a   What function do you think the rings or loops in the tracheae perform?

b   If there are ringed vessels in the midrib of a leaf, too, does this suggest they do a job similar to that of an insect's tracheae? Do the vessels in the plant appear to contain air?

c   How do tracheae end in the gut wall? Is there any evidence that they, like blood capillaries, eventually join larger tracheae to enable circulation of air in one direction?

d   If you find a trachea which appears to be discontinuous as in figure 13, does this give you any help in interpreting the 'blind' endings of the smallest tracheae?

e   Now that you have looked at the breathing movements of a living locust and the arrangement of tracheae inside a dead one, how do you think these animals bring about the ex-

Figure 12
Part of a tracheal system.

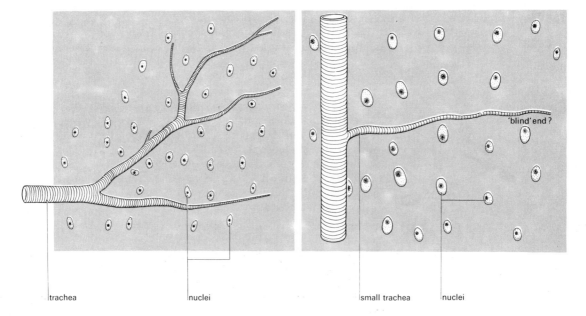

trachea   nuclei   small trachea   nuclei

Figure 13
Part of a tracheal system; it has been
immersed in alcohol for a long time.

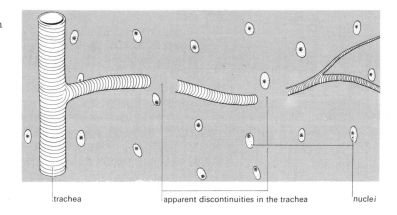

trachea   apparent discontinuities in the trachea   nuclei

change of gas between their internal organs and the surrounding air? Consider the size of the trachea.

2.4 Breathing apparatus of mammals

Next we should study our own breathing apparatus. Obviously, we are not in a position to dissect a human but as all
mammals have similar breathing organs, the lungs, we can
investigate a rat or mouse instead.

We know, from early childhood, that we can breathe air only
through our nose or mouth. If both are blocked or the neck
is constricted (strangled) the victim dies of asphyxia. We
differ greatly from insects. The aim of the following dissection is to trace the path of air from the nose and mouth into
the body, bearing in mind what we discovered when we dissected the locust. By starting at the abdomen and working
forward we can learn more than by immediately opening the
region of the chest.

To open the abdomen

Procedure

1 Put a freshly-killed mouse on its back, on the wax bottom
of a dissecting dish. Fix it by pins through the legs. Pull
these before pinning them, so that the mouse is firmly held.
Do not cover it with water.

2 Take hold of the skin in the middle of the abdomen with a
forceps, pull it upwards, and cut it with scissors, thus making a hole in the skin. From this hole cut forward with scissors right to the jaw and cut backwards towards the tail as
shown in figure 14. Separate the skin from the muscle under-
neath and pin it back or cut it away.

3 Carefully make a hole in the muscular wall at about the
same place as before (2). Cut forward until you come to the
flap of cartilage at the end of the sternum. Cut along the line
of the lowest ribs, left and right, as in figure 14. By making

further cuts, expose all the contents of the abdomen.

4   Identify the major organs of the abdomen, that is, the alimentary canal, the liver, the kidneys, and the reproductive organs. See what their relationship is to each other. You will find it extremely helpful to use a binocular microscope.

5   Find the oesophagus where it joins the stomach, tie cotton round it tightly, and then cut the oesophagus in front of this ligature so that the contents of the stomach do not escape. Pull the stomach away gently.

6   Remove the liver without damaging the diaphragm, the sheet of muscle separating the abdomen and the thorax. The liver contains a great deal of blood; wash away any which obscures the diaphragm.

7   Examine the diaphragm and test its tension by pressing it with a smooth glass rod or any small blunt instrument. (See the questions following these instructions.)

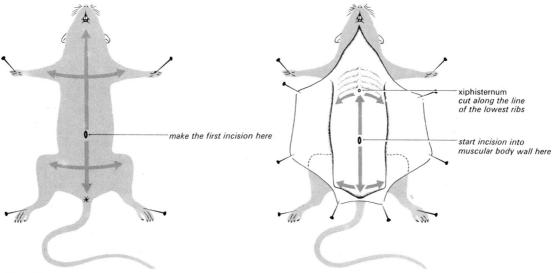

make the first incision here

xiphisternum
cut along the line
of the lowest ribs

start incision into
muscular body wall here

Figure 14
Dissecting the mouse – opening the abdomen.

To open the thorax

8   Feel the ribs with a seeker and carefully make a hole between two of them, well over to one side (figure 15).
Immediately check the tension of the diaphragm on the same side as the hole and on the other side.

9   Carefully insert a blade of your scissors into the hole and cut through the ribs along the lines indicated in figure 15, so as to remove a flap of tissue, revealing the contents of the thorax. Identify the lungs, the heart, and the vessels associated with them.

10 Extend the opening of the thorax forward by dissecting the neck. Look for two tubes, the oesophagus (gullet) and the trachea (windpipe). Separate the two from each other,

using a blunt seeker and a forceps. Remove the lower jaw and tongue; explore the nasal cavities in the head to trace the air passages as completely as possible.

11   Trace the trachea back towards the lungs. Record its position by sketches showing its relationship to the associated organs. When you have done this, remove the heart, wash out the blood, and set the heart aside in alcohol (50 per cent aqueous) for use later when you are studying the circulatory system.

Further investigation of the diaphragm

12   For this work a mouse is too small and it is better to use a freshly killed rat. Open the abdomen as described in stages (1)–(7). Cut into the neck and find the trachea. Expose about 1 cm, cut it transversely, and insert a metal cannula or a thin glass tube such as the one used as an extended jet on the burette for titration (figure 1). Tie cotton firmly round the trachea to make an airtight joint.

13   Connect the cannula or tube to a U-tube (bore 1–3 mm) containing coloured oil. Mark the levels of the fluid and then pull the diaphragm back, away from the head. Note any changes observed in the fluid levels. Push the diaphragm towards the head and again observe the fluid in the U-tube.

14   Open the thorax by making a small hole (8) and repeat the movements of the diaphragm (13); again, observe the levels of the fluid.

To investigate the physical nature of the lungs

Continue to open the thorax of the rat as in (9), (10), and (11) and remove the heart. Wash this and preserve it in alcohol. Disconnect the U-tube and attach a large syringe barrel to the trachea. Insert the piston and push it home to inflate the lungs. Note the volume of air you have used. Remove the syringe suddenly and note any change in the shape or volume of the lungs.

Figure 15
Dissecting the rat – opening the thorax.

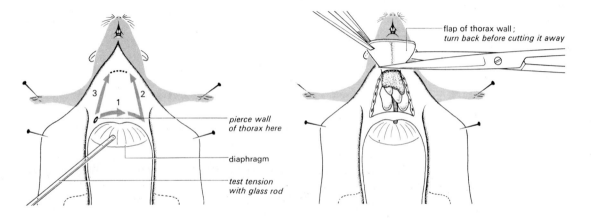

pierce wall
of thorax here

diaphragm

*test tension
with glass rod*

flap of thorax wall ;
turn back before cutting it away

Questions

a   Living muscle can contract and become taut. At other times it can be relaxed, and hardly tense at all. Do the muscles of the dead mouse appear to be contracted or relaxed?

b   The diaphragm is muscle. Was it tight or loose when you first examined it? Is the state of the diaphragm likely to be due to muscle contraction?

c   Describe the tension of the diaphragm when the thorax has been opened. If the tension has changed how do you account for this?

d   How can you distinguish between the oesophagus and the trachea; is the difference between them related to their functions?

e   Is there any similarity between the trachea of a mouse and a trachea in a locust?

f   If gases pass from the outside air to the lungs, how do they pass on to all the other tissues of the body? Try to put your answer in the form of a simple hypothesis based on your mammalian dissection rather than on previous knowledge.

g   A rise in the level of the fluid in the open side of the U-tube indicates a rise in pressure. When you moved the diaphragm, did you cause any changes in pressure in the trachea? What does your answer to this question suggest to you about the organs we use to breathe and the way they work?

h   Was there a change in pressure after the thorax had been opened? What does your answer imply or suggest?

i   What volume of air did you use to inflate the lungs of the rat to their full capacity? How does this compare with an estimate of the volume of the thorax in which they are normally contained?

j   Give a word that describes one physical quality of lungs that you saw when you suddenly removed the syringe.

2.5 The fine structure of the lungs

For the next step, you will need a microscope. This is to find out how gases reach the tissues of the body of a mammal and how they leave it. Apparently, they are not conducted directly, as in locusts, but only as far as the lungs. We shall be considering this in Chapters 3 and 4 also.

It is far more difficult to examine lung tissue under a microscope than to examine the tracheal system of an insect. Instead of being a thin sheet of cells like the gut wall, a lung is spongy. Also, structures in the lung do not stand out clearly, as locust tracheae do, until they have been treated with dyes. Therefore, before we can examine lung tissue it must be taken through three processes:

1. Fixing – immersing it in a liquid which preserves it and makes it hard enough to be cut thinly.

2. Cutting – embedding the tissue or organ in hard wax and then cutting it automatically on a microtome into very thin slices only a few microns thick.

3. Staining – it is usual to do this with two or more stains of the type which colours different kinds of cells, or parts of cells, specifically.

When you are trying to interpret a prepared slide you will obviously want to know what the stains are and the kinds of cells they colour.

You need to know a good deal and to be able to use your imagination to interpret a thin slice of tissue under the microscope. For example, suppose you are dealing with a conical structure. Perhaps you have seen sections of a cone made at different angles in mathematics classes. If so, you will know that a slice through this tissue might appear as a circle, ellipse, parabola, hyperbola, or triangle.

In fact, cones are rare in animal tissue but cylinders, particularly hollow ones (vessels), are common. Some shapes produced by section are shown in figure 16.

When you are examining a prepared section you must make a continuous effort to build a three-dimensional picture in your mind, from the two-dimensional section. This is made easier if you have a complete series of sections, cut one after the other, to hand, but you are unlikely to do so in school laboratories.

Because of the considerable difficulties we have outlined, we shall restrict the questions about lung sections to a few simple ones.

From dissecting the mouse and rat we can predict that air passages and blood vessels are present in the lungs. What we want to know is how they are brought close to each other.

Procedure

1 Examine under L.P. a prepared section, stained with Azan. This particular method colours nuclear material red, cytoplasm pink, and connective tissue blue. Search for tubes containing blood (corpuscles) and others which appear empty.

2 Find and examine air tubes (small bronchi) and blood vessels under H.P. Distinguish between blood corpuscles and cell nuclei (figure 17). Compare the linings of the two types of tube so that you could distinguish between them even if there were no blood corpuscles in one. Linings or coverings (epithelia) are most important in living animals, whether they are on the outside (skin) or on internal surfaces. The epithelia of bronchi are exposed to air and those of arteries

straight tubes

cross section ⟶

oblique section ⟶

longitudinal sections ⟶

⟶

⟶

curved tubes

⟶

⟶

⟶

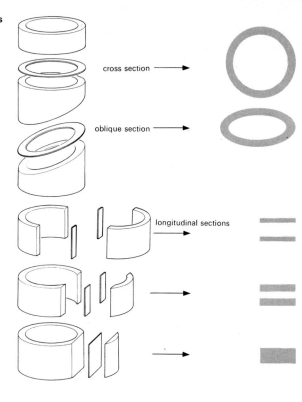
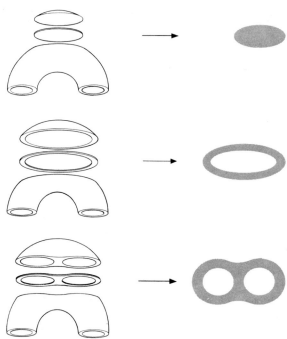

Figure 16
Sections of tubes cut in different planes.

and veins are exposed to blood. Note the differences, bearing this in mind.

3 Look at the spongy, diffuse part of the section, which occupies most of it. Search for consistent patterns in the apparently haphazard arrangement of the tissues. Using a micrometer eyepiece, estimate the average diameter of a space and also measure the thickness of tissue separating spaces.

4 Compare the prepared section with the photomicrograph (figure 18) of a section of a lung. In this preparation the blood vessels have been filled with a dye. A network of capillaries shows up clearly; compare it with the shapes of the spaces on your slide. Figure 19 will be helpful at this point.

5 Look at a slide of a lung section that has been stained to show up the elastic fibres. These are non-living structures made by cells. Note how these fibres are distributed.

Questions

a How many different kinds of tubes are present in the prepared slide of a lung section?

b What are the distinctive features of the tube linings?

c Are there any signs of rings or hoops, as there were in locust tracheae? (Remember you have been examining a section, not complete tissue.)

d Your lungs produce mucus which is sometimes brought up the trachea. Can you see cells in the bronchial epithelia which might be responsible for secreting mucus and for moving it? You can only guess here, as there is no way of examining living tissue.

e Comment on the distribution of elastic fibres, bearing in mind how a whole lung behaves after it has been inflated.

f From your measurements and estimates work out how far a gas molecule would have to travel from the centre of an

Figure 17
A small portion of lung section.

both corpuscles and nuclei appear red in Azan stain

blood corpuscles: note they are surrounded by empty space

nuclei: note that these are situated in cytoplasm (pink)

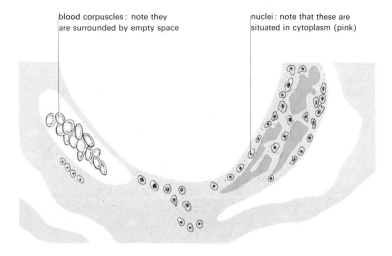

average tissue space to the side, and from the side of a space through the wall into a blood vessel.
Draw an air sac to scale.

g   Review the breathing systems of locusts (insect) and mice (mammal). Air, as such, goes only as far as the lungs of a mammal. Blood carries oxygen and carbon dioxide throughout the body, as many of you will know. Remember when you compare this system with the tracheal one of insects that locusts are large insects, but even the largest insect is minute compared with the largest mammals.

h   Why do you think that breathing systems exist at all? If the outer surface of the body were permeable to oxygen and carbon dioxide there would be no need for tracheal systems or lungs.

2.6 The capacity of the human lungs

Figure 18
Micrograph of the capillary network of lung tissues (×150).
Photo, B. Bracegirdle.

Now that we know something about the anatomy and physical properties of dead lungs we can think about how living

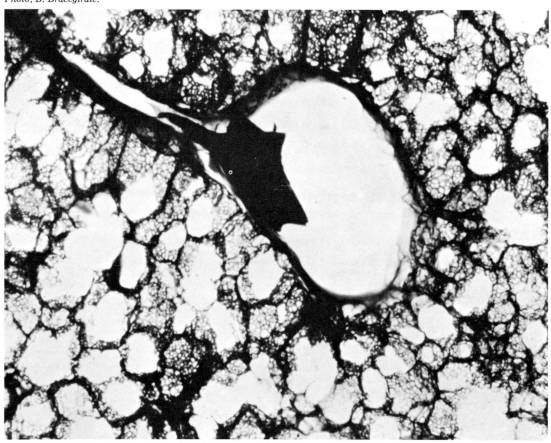

mammals such as ourselves exchange gases. We can use the spirometer to find the capacity of human lungs and to increase our understanding of the way in which the exchange of gases takes place.

Procedure

1   Fill the chamber of the spirometer with oxygen and set up the apparatus, including the kymograph, exactly as before (1.6). Take *all* the precautions you took then.

2   Let a subject, rested and sitting down, breathe into the mouthpiece. The sixth time he breathes out he should go on doing so past the normal point until he has forced out as much gas from his lungs as he is able. Take off the mouthpiece.

3   Start again, when breathing has returned to its normal rate. After he has breathed in and out normally six times with the mouthpiece on, he should start a forced inward breath and inhale as much oxygen as he is capable of doing. Immediately after doing this he should force his breath out again as far as possible.

4   Calibrate the spirometer if you have not already done so. We suggested a way of injecting a known volume of gas into the chamber in Chapter 1 (figure 6). If the spirometer pen moves upwards as the subject breathes in and draws gas out of the chamber, then you will get traces like those in figure 20. As you will have noted when dissecting dead mammals, lungs are never empty of air. If you doubt this, drop a portion of lung into water. The lungs or a cut-off portion of the lungs of an animal that has taken its first breath will float. The fact that leaves float has already been used in this chapter

Figure 19
A guide to the interpretation of lung tissue.
Based on Freeman, W. H. and Bracegirdle, B. (1967) An atlas of histology, 2nd edition, Heinemann.

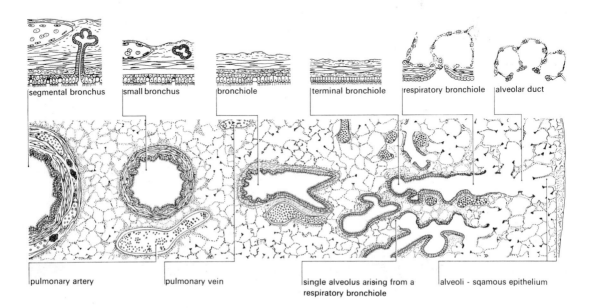

segmental bronchus   small bronchus   bronchiole   terminal bronchiole   respiratory bronchiole   alveolar duct

pulmonary artery   pulmonary vein   single alveolus arising from a respiratory bronchiole   alveoli - sqamous epithelium

to help us to study their structure. The air which even the hardest effort of breathing out cannot expel from the lungs is called *residual volume*.

5   Find the weight of the subject, in pounds.

Questions

a   What are the following volumes for the subject: tidal volume, expiratory reserve volume, inspiratory reserve volume? (See figure 20.)

b   What is the *vital capacity*, i.e. the sum of the three volumes discovered in (a)?

c   You can estimate the total lung capacity by multiplying the expiratory reserve volume by six. This has been found to provide a reasonably accurate figure. Work out the total lung capacity of the subject.

d   Calculate from (b) and (c) the residual volume.

e   Calculate the volume of air left in the lungs at the end of a normal expiration. Look at figure 20.
A normal inspiration brings in the tidal volume of air and this mingles with the volume calculated in (e). However, it is misleading to think that the tidal volume is completely available for gas exchange. Some of it occupies the trachea and other tubes of the system. This is called dead space and it can be very roughly estimated by converting the subject's weight in pounds into a volume in cm^3.* Thus, if a subject weighs 140 lb, the dead space of his trachea, bronchi, etc. is about 140 cm^3.

f   Calculate the dead space of your experimental subject.

g   Deduct this volume from the tidal volume and express the result as a percentage of the volume of air already present in the lungs (expiratory reserve plus residual volume).

h   Comment on the statement, 'breathing brings fresh air into close proximity with blood capillaries in the lungs', in the light of your calculations. Summarize the physical phenomena and bodily activities responsible for gas exchange.

Figure 20
A spirometer record showing lung volume.

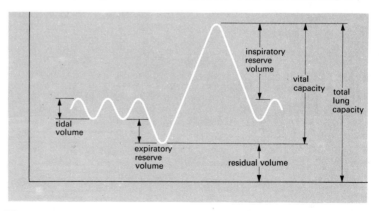

*There is no simple equivalence using metric weight; 0·45 kilogramme is roughly equivalent to 1 cm^3 dead space.

Bibliography

Best, C. H. and Taylor, N. (1964) *The living body*. 4th edition. Chapman & Hall. (Mammalian and human breathing systems.)

Clegg, A. G. and Clegg P. C. (1962) *Biology of the mammal*. Heinemann. (Mammalian and human breathing systems.)

Fogg, G. E. (1963) *The growth of plants*. Penguin. (Leaf structure and the function of stomata.)*

Rowett, H. G. Q. (1960) *The rat as a small mammal*. 2nd edition. John Murray. (Mammalian and human breathing systems.)

Simon, E. W., Dormer, K. J., and Hartshorne, J. N. (1966) *Lowson's botany*. 14th edition. University Tutorial Press. (Leaf structure and the function of stomata.)*

Thomas, G. (1964) *Dissection of the locust*. Witherby. (Locust breathing system.)

*Most botany textbooks also provide this information.

Synopsis

1 Oxygen is obtained from the environment by insects through tracheal systems, by mammals through lungs, and by fishes through gills. These organs are best investigated by dissection. The dogfish is a suitable animal for the investigation of a gill system.

2 Dissection reveals macroscopic structures; to follow the path of oxygen absorbed from the surrounding water, it is necessary to examine prepared sections of gill tissue under the microscope.

3 In the time available, full knowledge of the ways in which gills work cannot be acquired by investigating their anatomy alone. Some results from experimental work are introduced as additional evidence.

4 The way in which oxygen passes from the environment into the body of a dogfish is investigated elsewhere (section E.2).

Gas exchange systems: the gill system of the dogfish

Summary of practical work

section   *topic*

E1.1   External features and preliminary dissection

E1.2   The anatomy of a branchial bar

E1.3   Histology of the gill lamellae

E1.4   The mechanism of ventilation

E1.1 External features and preliminary dissection

The way in which an animal functions cannot be fully understood without some knowledge of its internal anatomy. The method by which this is investigated is dissection. In the example which follows, the overall aim is to discover how the structure of the gills is related to their functions. We know that in order to live, fish need the oxygen dissolved in water and gills play some part in obtaining it.

The dogfish, like sharks and other cartilaginous fish, has a gill system which opens directly into the surrounding water. Bony fish, such as the herring and cod, have a more compact system covered by a flap of tissue. Dogfish are therefore a little easier to dissect. The terminology used below is the same as that in standard texts, to which you should refer if you need to clarify a term. Look at the questions in advance and consider the steps you would take if no instructions were given.

Procedure

1   When carmine particles are placed at the external openings of the spiracles of a live dogfish they emerge from the first three pairs of external clefts. Placed near the mouth, they

emerge from the last three pairs. Examine the mouth, the spiracles, and the septa guarding the external branchial clefts.

2 Separate the septa of the external clefts and observe the branchial pouches and the gill lamellae lining them.

3 Make a deep incision on the right side of the animal from the angle of the jaw to the posterior limit of the pectoral fin, dorsal to the fin base. This cut must go through to the cavity of the gut; try to cut through the middle of each branchial bar (see figure 21).

4 From the posterior limit of the first incision make a transverse cut right across the animal ventrally, to the left pectoral fin (see figure 22). (The terms left and right apply to the animal's left and right sides; if the dorsal surface of the fish is against the board, the animal's right will be on your left.)

5 Use two awls to pin out the right side of the animal, placing them well clear of the branchial pouches. Make sure that they are firmly inserted into the dissecting board.

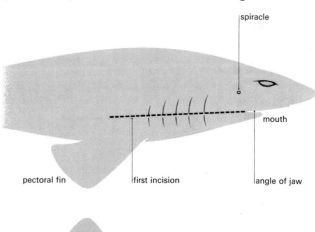

Figure 21
Diagram to show the first incision.

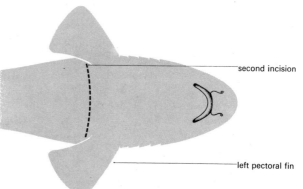

Figure 22
Diagram to show the second incision.

6   Pull the freed lower jaw and ventral part of the pharynx over to *your* right. To free this region completely you will have to cut through the oesophagus and the liver. Fix the deflected part firmly to the board with two awls – one through the lower jaw and one through the right pectoral fin. The cut anterior parts of the liver go over with the deflected region (see figure 23).

7   Wash out the contents of the gut and clean out the branchial pouches.

Questions

a   From your examination (stage 1), which structural features do you think could act as valves controlling the direction of water as shown by the carmine particles?

b   Describe the distribution of lamellae in each of the pouches of one side.

c   Make a simple diagram of the pharynx showing the direction of flow of water through it.

d   How do you think the fish propels the water along the paths indicated by your diagram? Relate your hypothesis to observations made so far in this dissection.

E1.2 The anatomy of a branchial bar

Having revealed the inside of the pharynx and seen something of the gill surface, it is now necessary to observe these in more detail by removing part of a branchial bar.

Procedure

1   Remove a small portion of a branchial bar which has already been cut in stage 3 of the previous section.

2   Examine it carefully with reference to figure 25 and remember how it was orientated in the whole fish.

Figure 23
The pharynx displayed.

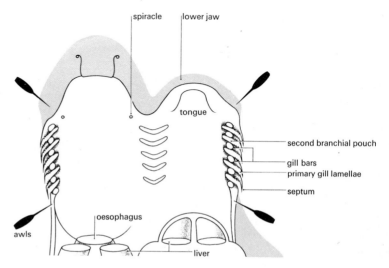

3   Dissect the portion of branchial bar to discover the distribution of muscle and skeletal (cartilaginous) components.

Questions

a   Describe the distribution of primary lamellae on both surfaces of the cut portion of branchial bar.
b   How are the secondary lamellae arranged with respect to the
primaries?
c   Describe the skeletal components in a branchial bar. Do any
of them extend into the septum? (You would find it helpful
to inspect a preserved and mounted dogfish skeleton.)
d   What muscle tissue can you find in the portion of branchial
bar? What do you think happens in the living fish when this
muscle contracts?
e   By what possible pathways can oxygen, absorbed from the
surrounding water, pass from the branchial bar to other
parts of the body?

E1.3 Histology of the gill lamellae

The third stage of this investigation requires thin sections
of gill tissue cut with a microtome, stained, and mounted. It
is essential that these sections should be related to the branchial bar, already examined, since the overall aim is to see
how this structure is adapted to its function.

Procedure

1   Examine the sections under L.P. and identify the anterior
and posterior surfaces. Alternatively, examine photomicrographs of gill sections.
2   Using any microscope sections available, revise the appearance of cartilage and muscle. Return to the gill slide and
distinguish between any muscle, cartilage, and blood vessels
present. Try also to identify the primary and secondary
lamellae.

Figure 24
Diagram of the pharynx of a dogfish.

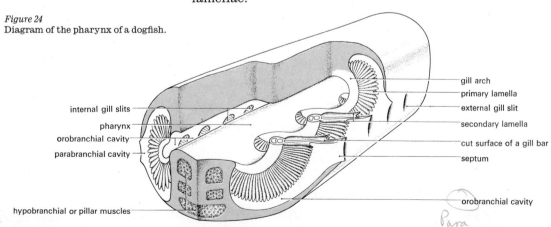

internal gill slits
pharynx
orobranchial cavity
parabranchial cavity

hypobranchial or pillar muscles

gill arch
primary lamella
external gill slit
secondary lamella
cut surface of a gill bar
septum

orobranchial cavity

3 Examine the lamellae under medium and high power magnifications.

4 Use cardboard, Plasticine, or modelling wax to make a model of part of a branchial bar to show the arrangement of the lamellae, the septum, and, if possible, the skeletal support of the bar.

Questions

a In what respects do the lamellae, as seen in section, meet the specification for an 'ideal absorptive organ'?

b If oxygen is absorbed by the lamellae from the environment, how does it pass from them to the branchial bar? Give your reasons, based on observations of gill sections under the microscope.

c What mechanism can you suggest for the ventilation of the secondary lamellae?

d What features in common have the gill lamellae of a fish and the air sacs of a mammal's lungs? Comment on the significance of these common features.

E1.4 The mechanism of ventilation

In a living fish the gills are perpetually ventilated with water. To a large extent, you will be able to determine the way in which the pharynx brings this ventilation about, from your previous observations. But you need some additional information from experiments on living fish before you can fully deduce the mechanism. This is included in the following questions, which form the basis of a coherent argument.

Questions

a Ventilation of the gills involves a cycle of events. As water enters and leaves, the pharynx appears to get larger and smaller. During the expansion phase, when water is entering the pharynx, what happens to the orobranchial volume and pressure? (Refer to figure 24 for the meaning of the terminology.)

Figure 25
Diagram of a piece of a branchial bar of a dogfish.

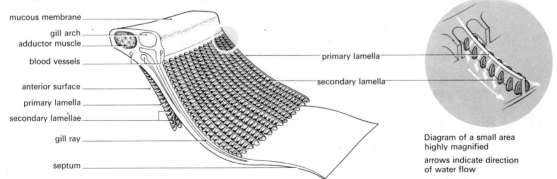

mucous membrane
gill arch
adductor muscle
blood vessels
anterior surface
primary lamella
secondary lamellae
gill ray
septum

primary lamella
secondary lamella

Diagram of a small area highly magnified

arrows indicate direction of water flow

b Now consider the cartilaginous branchial rays that exert an outward force. What will happen to the volume and pressure in the parabranchial cavity due to this force?

c Are the two cavities in free communication or is there a barrier that would delay pressures being equalized should they differ? What anatomical structures could form a barrier between the orobranchial cavities?

d What would happen to the water in the pharynx if there were a difference in pressure between the two cavities?

e Measurements made on anaesthetized dogfish show that parabranchial pressure falls below that of the orobranchial cavity during the expansion phase.
1. What effect will this have on the flow of water?
2. Can this water escape from the fish?
3. If not where will it be driven?
4. Does this in any way serve to ventilate the respiratory surface?

f *1.* When the contraction phase begins will pressures be equal?
2. If not, which cavity will have the greater pressure?

g Due to this difference in pressure, what will happen to the flow of water?

h *1.* How many pumps operate to ventilate the respiratory surface?
2. What kind of pumps are they – pressure or suction pumps?
3. Do they act independently or are they synchronized?
4. Is the flow of water over the gill lamellae continuous or intermittent?
5. Is the ejection of water from the external clefts continuous or intermittent?

i Summarize your answers (a) to (h) by writing a short account describing how the gill lamellae of the dogfish are ventilated.

Bibliography

Grove, A. J. and Newell, G. E. (1966) *Animal biology*. 7th edition. University Tutorial Press. (A comparative treatment of vertebrate anatomy.)

Hughes, G. M. (1963) *Comparative physiology of vertebrate respiration*. Heinemann.

Hughes, G. M. (1961) 'How a fish extracts oxygen from water.' *New Scientist* 11, 346. (Experimental evidence is combined with anatomy to provide explanations of gill function.)

Wells, T. A. G. (1961) *Three vertebrates*. 2nd edition. Heinemann. (Comparative account of structure and function.)

Synopsis

1 This chapter considers how the materials which organisms exchange with their environment circulate within animal and plant bodies.

2 We can easily examine leaves for signs of internal activity and movement. It is more difficult, but perhaps more rewarding, to study the circulation of blood in a living animal.

3 Circulation depends upon the action of a pump; a frog's heart provides a good subject for observation and experiment.

4 Investigating the structure of hearts and blood vessels adds to our understanding of circulation in animals.

5 We consider some experiments by other people on the movement of materials in the stems and leaves of plants and some hypotheses which attempt to explain it.

Chapter 3

Transport inside organisms

Summary of practical work

section *topic*

3.1   Movement inside plant cells

3.2   Circulation in animals

3.3   The heart in action

3.4   The structure of hearts

3.5   Blood vessels

3.6   Transport inside plants

In the last chapter we investigated three kinds of pathway through which the exchange of gas can take place.

Leaves contain many spaces in which gases are free to diffuse; there are no pumps to drive air in and out. Locusts have a system of tubes (tracheae) leading from the body surface to all internal organs, but in rats and mice gas exchange is localized in special organs and these animals rely on blood to carry oxygen from the lungs to all parts of the body.

Blood carries more than just dissolved gases, and because insects have tracheal systems for gas exchange, it does not mean that they necessarily lack circulating blood. Exchange of materials between the environment and internal tissues implies the movement of molecules. There are two ways in which atoms and molecules can move about, illustrated in figure 26:
a. by mass flow, when they are all moved along together.
b. by individual random motion.

For mass flow to occur from X to Y as in figure 26, the pressure must be greater at X than Y. By contrast, random motion will cause some molecules to diffuse from XX to YY even though there is no difference in pressure. An approximately equal number will move in the opposite direction, from YY to XX. Molecules are too small to be seen under a microscope but the movements of microscopic particles suspended in liquid reveal something of the activity of the molecules in the liquid. In mass flow, the particles are swept along together; in still liquids small particles are agitated by the random motion of molecules. This agitation of small, visible particles by minute, invisible molecules is called Brownian movement.

3.1 Movement inside plant cells

With these ideas in mind we can search in animals and plants for signs of random motion, mass flow, and circulation. Let us consider plant cells first.

Procedure

1   The leaves of the pond weed *Elodea* are thin compared with those of land plants. Cut one off and mount it in a drop of pond water on a slide. Cover with a cover-glass.
2   Observe under L.P. and H.P.

Questions

a   Can you observe any moving particles? If so, is the movement from cell to cell or confined within cells?
b   What functions do you think these movements could be performing?

Movement inside animals

The breathing systems of locusts depend on the diffusion of gases. This is aided by mass flow produced from pumping movements. In mammals breathing is localized in the breathing organs – the lungs – and the diffusion of dissolved

Figure 26
Diagrams representing the two ways in which molecules move.

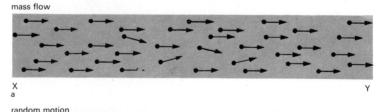

mass flow

X
a

Y

random motion

XX
b

YY

leaf

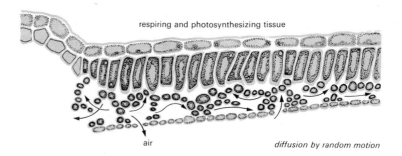

respiring and photosynthesizing tissue

air

diffusion by random motion

insect

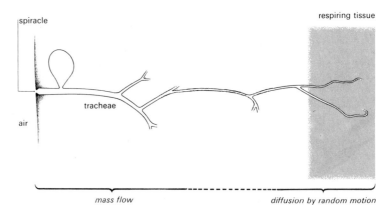

spiracle

respiring tissue

tracheae

air

mass flow   *diffusion by random motion*

mammal

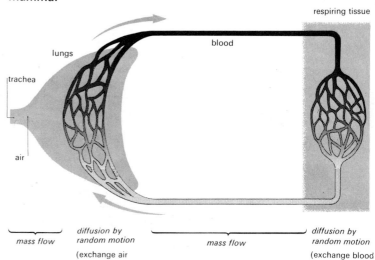

respiring tissue

lungs

blood

trachea

air

mass flow   *diffusion by random motion* (exchange air | blood)   *mass flow*   *diffusion by random motion* (exchange blood | tissue cells)

Figure 27
Diagram comparing three examples of
molecular movement.

gases to all parts of the body would be far too slow to supply the tissues adequately. The exchange of materials such as oxygen and carbon dioxide depends on the mass flow of air and of the blood in which the materials are dissolved. (See figure 27.)

The speed of diffusing molecules depends on their state, whether this is gas, liquid, or solid, and on temperature. The hotter a fluid is the faster its constituent molecules travel, and vice versa. The temperatures of living organisms seldom rise above 40° C. A low rate of diffusion impedes the movements of materials. Therefore only in small organisms or thin tissues is diffusion an adequate means of supplying oxygen and nutrients to the innermost cells. In larger organisms and tissues diffusion is aided by mass flow.

Mass flow in an organism can be achieved in two ways (figure 28): by the intake of material at one point and output at the same or another point; by circulation.

Figure 28
Three examples of mass flow.

two-way flow (air to the lungs)

one-way flow (water through a plant)

circulation of blood

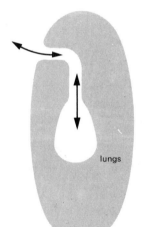

lungs

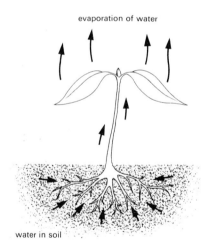

evaporation of water

water in soil

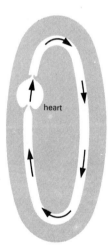

heart

3.2 Circulation in animals

To maintain the circulation of a fluid in any system, living or otherwise, there are three requirements:
Vessels to contain the fluid – these need not be tubes.
A pump of some kind to cause changes in pressure.
Valves to allow movement in one direction only.

The fluid which is circulated through animals is usually called blood but its composition and appearance differ greatly throughout the animal kingdom. We shall consider this in greater detail in the next chapter.

We can observe the movement of blood in vessels, for example, in the thin skin between the toes of frogs and in the tails of tadpoles, particularly those of the toad *Xenopus* because they are transparent. Continual movement in one direction implies circulation. More striking examples are found in the external gills of young tadpoles of the common frog *Rana temporaria* and in some fresh water crustaceans such as the water louse *Asellus*. Water lice flourish in laboratory tanks (discussed as containers for model communities in Chapter 1). They are fairly transparent and therefore particularly suitable for study.

Procedure A General circulation

1   Capture a number of water lice from an aquarium tank. The best for this work are those from 5 to 7 mm long which are light in colour. Larger and darker ones may be useful for a practice 'run'.
2   Mount one living *Asellus* on its back in a generous drop of water in the manner shown in figure 29. Adjust the slide so that the animal is trapped but not squashed. Examine it under a microscope.
3   Observe any limb under L.P. and search for the movement of materials inside it. The tapering ends of limbs are the most transparent parts, and even in well-pigmented water lice the antennae are usually sufficiently transparent to allow their contents to be seen.

Questions A

a   Can you see any moving objects inside any of the limbs? Record your observations accurately with the aid of simple sketches.
b   Do the particles or corpuscles within a limb all move in the same direction?
c   Can you see blood vessels in any of the limbs or antennae?
d   Is the rate of flow of corpuscles regular?

Figure 29
A method of mounting an *Asellus*.

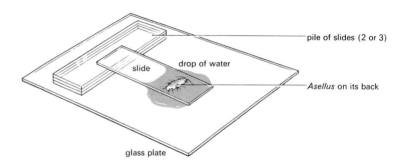

The sight of circulating material prompts the question: 'What is being transported?' Water lice live in water and remain beneath the surface. Presumably they need oxygen, like all animals, and obtain it in dissolved form from the water in which they live. Does this oxygen diffuse through all parts of the body's surface? This is a very difficult question to answer. These animals are too large to rely solely on simple diffusion and it is therefore reasonable to assume, as a hypothesis, that the blood of water lice carries oxygen. Rats, mice, and other mammals have lungs, fishes have gills; do water lice too have localized organs for obtaining oxygen? If you put an *Asellus* in a test-tube or specimen tube of water and look at it from one side, you will see that there are always movement and agitation near the rear end of the animal. This is so even when the animal is stationary. This is suggestive, and leads us to continue our investigation, concentrating on the rear part of *Asellus*.

Procedure B Circulation through gills	4	Observe the posterior part of the trapped *Asellus* under L.P. Press the covering slide down gently and firmly, but with extreme care. Watch for the appearance of part of a colourless structure on one side or the other. Figure 30 illustrates the place to watch and the proportions of the structure relative to the whole body.
	5	If no such structure appears, move the pile of slides to the left (see figure 29) and apply a little more pressure. If this brings no success repeat the procedure with another *Asellus*.
	6	When a colourless structure appears observe it under L.P.
Questions B	e	Suggest one reason why the structure appears colourless in contrast to other parts of the body.
	f	Can you see particles moving inside this structure and, if so, describe the motion and its direction in relation to the whole body?

Figure 30
Ventral view of posterior part of *Asellus*.

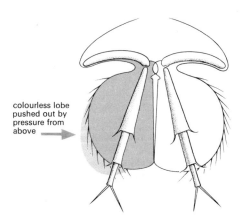

colourless lobe
pushed out by
pressure from
above

g How does this direction compare with that of water outside the body when you observe it in the specimen tube?

h Structures of this kind are termed 'gills' by the writers of textbooks. What evidence have you obtained to support the idea that they do, in fact, function as gills?

3.3 The heart in action

Though the movement of particles, and presumably blood, is clearly visible in parts of *Asellus*, it is usually quite impossible to make out what is driving the blood. If there is a heart it lies in the thickest and least transparent part of the body. There may be one or a number of 'pumps' or hearts at work. You can see a single heart in action in smaller crustaceans such as the water flea, *Daphnia*. It is easy and rewarding to examine it under the microscope.

To investigate the action of a working heart and conduct experiments on it, we need a larger animal. Because it is not transparent the body must be opened to show the heart. It is possible to kill some animals, such as frogs and toads, in a way that leaves the heart still functioning. This is done by taking an unconscious frog or toad and destroying the brain and spinal cord. The operation is called 'pithing' and requires skill and training if it is to be carried out successfully. You must not attempt it.

Experiments on the heart call for team work and division of labour so that several things are ready and working at one time. The procedure which follows is subdivided into parts A, B, and C so that one student or a group can each be responsible for one part.

You should not assume that the movement of blood through the tissues of a frog is the same as through a water louse. When you can, observe the thin skin between the toes of a pithed frog, under the microscope. Compare the movements of corpuscles with those seen in *Asellus*.

You will require a kymograph, paper, heart lever, frog board, pithed frog, pins, clean dissecting instruments, cotton thread, and amphibian Ringer's solution.

Procedure A Preparation of the kymograph

1 Remove the kymograph drum.
2 Fix a sheet of graduated or glazed and smoked paper to it.

Procedure B Assembling the heart lever

3 There are several forms of lever. Find out how yours is supposed to work. Commonly, one uses a light pivoted lever with a pen or scraper fixed at the end. Assemble the parts and clamp the lever onto its stand. (See figure 33.)

4   Bend a small pin in a rough figure S to make a hook. Attach a length of cotton (20 cm). (If a heart clip is provided, there is no need to make a hook.)

Procedure C Dissection of the frog

5   Fix a pithed frog on its back on the frog board by pins through the feet. Work carefully but as quickly as possible.

6   With fine scissors puncture the skin in the centre of the abdomen and cut all the way forward to the lower jaw. Make further cuts from the first one, along the mid-line of the limbs. Fold back the skin.

7   Note the dark red line of the anterior abdominal vein situated centrally in the muscle of the abdomen. Cut through this muscle, parallel with the vein and a few millimetres from it. Continue forward until you meet the xiphoid cartilage. (See figure 31.)

8   If possible, look at an articulated skeleton of a frog as well as at figure 31. Cut as shown by the dotted lines, with strong scissors, severing the coracoid and clavicle bones on *both* sides of the mid-line. Avoid any downward movement of the scissors which is likely to damage underlying organs.

9   Hold the xiphoid cartilage with a pair of forceps and pull it upwards, together with the xiphisternum and central portion of the cut bones. When the whole flap has been hinged back, cut it off as shown (figure 31). Pull the forelegs apart and pin them down again.

10   Identify the heart; it should still be beating. Take hold of the thin membrane (pericardium – see figure 32) which surrounds the heart, cut a small hole in it, and remove it all. You will have to displace the heart upwards to complete this operation. Moisten the heart with Ringer's solution and

Figure 31
Dissection of a frog.

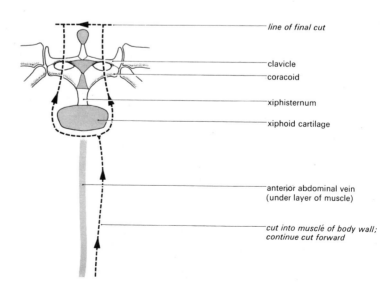

line of final cut

clavicle

coracoid

xiphisternum

xiphoid cartilage

anterior abdominal vein
(under layer of muscle)

cut into muscle of body wall;
continue cut forward

keep it moist in this way throughout the rest of the procedure. This solution is similar in composition to the fluid inside an unopened frog.

11   Insert the prepared hook (or attach a heart clip) into the ventricle of the heart (figure 33). Try to get the pin firmly into the heart at the first attempt, near but not too near the apex. The success of the experiments depends on the hook remaining firmly in position.

12   Align the frog under the heart lever, and attach the cotton to

Figure 32
Anatomy of frog heart and adjacent vessels, and position of the cardiac branch of the vagus nerve.
After Rowett, H. G. Q. (1967) Dissection guides I: The frog, *John Murray.*

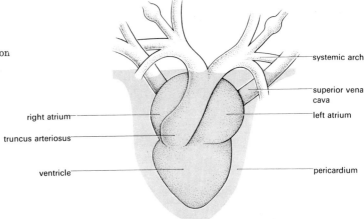

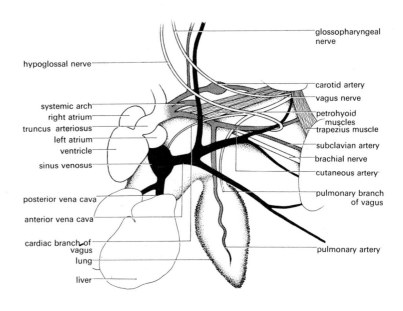

it so that each time the heart beats (contracts) the lever moves. Adjustments should be made so that this movement is as great as possible without the load adversely affecting the heart action.

13   Set the kymograph, with paper covered drum attached, at a slow speed, e.g. one revolution in three minutes, and move it into a position so that the pen or scraper of the heart lever makes a trace on the drum. After one revolution, move the kymograph away.

14   While the normal heart beat is being recorded (13) warm some Ringer's solution (1 dm³) to a temperature of about 30° C, place some in a refrigerator, and make ready solutions of adrenaline (1 part in 1000) and acetyl-choline (1 part in 1000).

15   Move the kymograph drum up or down so that a clear portion is ready for another trace. Repeat stage (13) but as soon as the lever begins to trace a line, direct a small but steady jet of warm Ringer's solution onto the heart. Let it continue for one revolution. Measure and record the temperature of the solution.

16   Similarly investigate the effects of cold Ringer's solution. Measure and record the temperature.

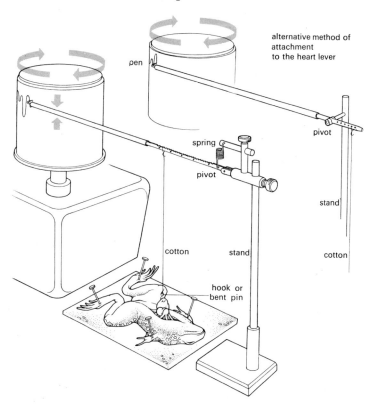

Figure 33
Frog prepared for experiments.

17   Make a record of heart beat after applying a solution of acetyl-choline. Wash the heart well with Ringer's solution and then apply adrenaline and again make a record.

18   Move the drum so that the lever writes a line along the bottom and move it with your hand once every second while the drum is revolving. This makes a time scale on the recordings, for future reference.

19   Finally, remove the paper and, if it is smoked, varnish or spray it with artist's fixative.

20   There are many experiments which can be performed on a frog's heart as well as those mentioned so far. When you have reached a satisfactory standard in preparing and dissecting the frog and recording your findings, you should attempt the following more difficult procedure because it provides important information about the way the heart works. Find the nerve connected to the heart; that is, find the cardiac branch of the vagus (figure 32). If you have an electrical stimulator apply repetitive pulses of electricity at maximum frequency through electrodes attached to the cardiac nerve. Record the heart beat as before. Increase the voltage of the stimulus until this affects the heart beat.

21   When you have again obtained normal heart beat, cut the cardiac nerve on one side. Mark the record to indicate the time when you do this. When you have completed the experiments remove the hook and pins and put the frogs in a suitable preservative so that they can be used later.

Questions

a   Examine the tracings made by normal heart beat before heat, and chemical or electrical treatment. Are the peaks simple, that is, V-shapes? If not, what kind of heart activity would account for the shapes obtained?

b   Which parts of a frog's heart contract and in what sequence?

c   What is the effect of temperature on heart rate?

d   What effect does acetyl-choline have on the heart?

e   What effect does adrenaline have on the heart?

f   What happens to the heart beat frequency when the vagus nerve is stimulated electrically?

g   Would you expect the cutting of the vagus nerve to have any effect on the heart? Give the reasons for your answer.

3.4 The structure of hearts

Now we have seen a heart contracting rhythmically, we will have to look more closely at its anatomy if we are to find out how it propels blood.

Procedure

1   Obtain the preserved hearts of mice and rats which you set aside earlier, together with the piece of locust body wall to which the heart is attached.

2   Examine the preserved frogs from the previous investigation

and note the many blood vessels arising from those connected to the heart. Identify the main viscera, that is, the lungs, liver, alimentary canal, and excretory and reproductive organs.

3 Cut out the heart together with short lengths of arteries and veins attached to it.

4 Obtain a fresh lamb's heart (from a butcher).

5 Examine the external features of all the hearts carefully as suggested by the questions below.

6 Investigate the anatomy of the hearts by cutting them into a series of sections. Consider carefully which is the best plane to make the parallel cuts in, so as to obtain the most information. To see sections of the heart of a mouse, use a binocular microscope. Make simple outline drawings.

Questions

a Describe the heart of a locust as far as possible from the preserved tissue at your disposal.

b What external features do the hearts of rats and sheep have in common which distinguish them from a frog's heart?

c Which hearts exhibit signs of possessing their own circulatory system?

d How does the total volume of the atria compare with the total volume of the ventricular space in any one heart?

e Is there any anatomical evidence that ventricles cause higher blood pressure than atria?

f How are valves prevented from 'blowing back' into atria when ventricles contract?

g Does your examination of heart sections suggest that the pressure produced by the two ventricles is equal or unequal? Give your reasons.

h What is the function of each of the chambers of the mammalian and the frog's heart? Base your answers on your own observations and information from textbooks. What is meant by 'double' circulation?

3.5 Blood vessels

The hearts of frogs, mice, and other vertebrate animals pump blood round the body through vessels. This can swiftly carry substances which diffuse into the body at one point to other regions before they diffuse out again into tissue cells. Vessels that are best suited to this kind of exchange, by diffusion, are not ideal for carrying quantities of fast-flowing blood. Small vessels with thin walls and a relatively large surface area are best for allowing molecules to pass into and out of the blood they contain but, at the same time, they present great resistance to flow. Moreover, they cannot withstand much fluid pressure. Figure 34 illustrates this.

It follows that in a simple circulatory system, consisting of

Figure 34
Diagram of longitudinal sections of
small and large blood vessels.

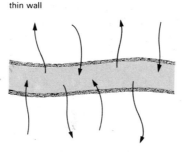

thin wall

a small tube (in section) suitable for
exchange; no molecules are far from the
walls of the tube

thick wall withstands high pressure

a large tube (in section) suitable for
conduction of a liquid; although there is
friction against the wall, liquid nearer the
centre can move rapidly

a pump (heart) and a network of fine tubes (capillaries) pre-
senting resistance to flow, the pressure at H, as shown in
figure 35, will be greater than at L. Bearing this in mind, we
can examine sections of blood vessels to see if their structure
corresponds to differences in the pressure of the blood they
contain.

Procedure

1   With the help of a hand lens, examine prepared sections of
three different blood vessels which have been stained to
show the presence of elastic fibres.
2   Refer to the questions below and examine the sections
further with a microscope. Remember that the vessels were
empty of blood when fixed and cut, so that some of the
features you see under the microscope may be due to the
method of preparation.
3   Examine the sections for elastic fibres. These appear as well
defined, thin, wavy lines. Draw three simple plans to show
the relative amounts and distribution of elastic fibres in the
three sections.

Questions

a   Which of the three vessels do you think contained blood at
high pressure? Give your reasons.
b   Which of these sections was taken from an artery nearest to
the heart? State your reasons.
c   What possible function can elastic tissue have in the wall of
a blood vessel? What type of blood vessel does it occur in?

3.6 Transport inside plants

Most animals have circulatory systems of some kind. Exceptions occur in a few groups of simple organisms, some with
bodies only a few millimetres thick. These manage to survive by unaided diffusion.

Some plants grow to enormous size; the majority have stems
and roots thicker than a few millimetres and extend many

centimetres in length. Presumably they too require a circulatory system. There is an implicit assumption here which
we must not overlook. The necessity for rapid transport
depends not only on the size of the organism but also on the
requirements of its tissues. If, for example, the internal
organs of locusts required little oxygen then the need for
rapid conduction and mass flow would not be great. We
should not assume that the requirements of plant tissues are
the same as those of animals.

Do materials move in plants by other means than simple
diffusion? You may have already observed the contents of
cells circulating in the leaves of *Elodea*. Now examine figure
36. A drop of sucrose solution with radioactive carbon atoms
(^{14}C) was applied to one leaf of each privet shoot, as shown.
After twenty-four hours the shoots were placed flat against
a photographic film in the dark which was subsequently
developed. Radioactive carbon affects film in the same way
as light; the developed film, called an autoradiograph, provides a record of the distribution of ^{14}C after twenty-four

Figure 35
Diagram of a simple circulatory
system.

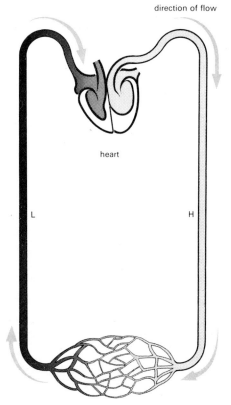

direction of flow

heart

L

H

capillary network

hours. One privet shoot had been 'ringed'; that is, the outer portion of the stem had been removed (see figure 37).

Questions

a   Describe the distribution of radioactive carbon in the control shoot after twenty-four hours, as shown in the autoradiograph. How does the distribution of ^{14}C differ in the two shoots?

b   Examine the micrograph of a privet section (figure 37). What can you deduce about the probable pathway of the radioactive carbon?

c   The total length of the shoot used was about 25 cm. What statements can you make about the speed of conduction of radioactive carbon in the living shoots?

In the middle of the last century, Graham carried out experiments on rates of diffusion. Putting a mixture of salt – NaCl – and calcium bicarbonate – $Ca(HCO_3)_2$ – at the bottom of a

Figure 36
Autoradiograph of privet shoots.
Photo, Dr J. W. Hannay.

control shoot A   auto–radiographs taken after 24 hours   experimental shoot B

cylinder of water, height 28 cm, he found that after six months the concentrations of the solutes at the top and bottom of the cylinder were unequal. The sodium chloride was in the ratio 11:12 and the calcium bicarbonate 1:4. In further experiments he compared the rates of diffusion of salt–NaCl–and glucose–$C_6H_{12}O_6$–and found them to be in the ratio of 100:36.

Plant cells contain protoplasm and can hardly be compared to a cylinder of water. If you pour copper sulphate solution onto a dish of jelly (gelatine) the blue colour can be seen to penetrate (diffuse) into the jelly to a depth of 1 cm in about twenty-four hours. (Much depends on the concentrations of copper sulphate and jelly, and on the temperature.) There is abundant evidence to show that diffusion in liquids is a slow process and that the larger the molecules concerned, the slower is their rate of diffusion.

The experiment with 'labelled' sucrose shows two things. Firstly, ^{14}C passes more rapidly through a plant than can be accounted for by simple diffusion alone. Secondly, this transport does not take place through the inner part of the stem (xylem). If you look at a prepared *longitudinal* section of a plant stem such as privet or sunflower, you will see that it is the xylem which provides continuous tubes which look

Figure 37
Transverse section of privet stem
($\times$100)
Photo, W. J. Garnett.

part removed in ringing

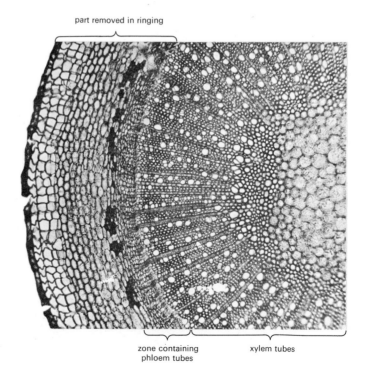

zone containing
phloem tubes

xylem tubes

Figure 38
Drawing of xylem and phloem cells –
longitudinal section.

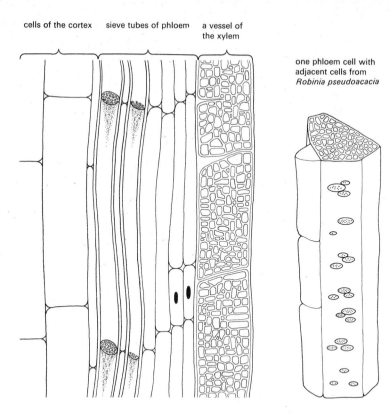

Figure 38
Drawing of xylem and phloem cells –
longitudinal section.

cells of the cortex   sieve tubes of phloem   a vessel of
the xylem

one phloem cell with
adjacent cells from
Robinia pseudoacacia

as though they could conduct liquids at high speeds. The
cells of all other tissues have transverse cross-walls of one
sort or another which appear to block the path of any moving liquid. Most interesting of these outer tissues are the
sieve tubes of the phloem. (See figure 38.) Because cross-
walls are fewer per unit of length in sieve tubes, it has long
been assumed that materials like sucrose pass up and down
in the phloem. The xylem (or wood) conducts water.

As long ago as 1923, H. H. Dixon measured the rate of accumu-
lation of carbohydrate in a potato tuber and calculated that
a rate of movement of 50 cm per hour would be necessary for
the conduction of carbohydrates (mostly as sucrose).
Though this was considered at that time to be impossibly
fast, more recent studies using ^{14}C have revealed rates from
50 to 100 cm per hour in various plant stems.

It may seem surprising that, although we know that materials travel at these considerable speeds in both directions, no
one has yet proposed a satisfactory hypothesis which
accounts for all the facts. Many hypotheses have been suggested and these can be broadly classified under two head-
ings: *Mass flow* – hypotheses involving the flow of sap

through sieve tubes; *activated diffusion* – that is, some process like diffusion but much faster.

Figure 39 summarizes, in diagrammatic form, some ideas which have been put forward, including a suggested means of circulation in plants. This is developed in figure 40 to show a system without mechanical pumps or valves. When a satisfactory hypothesis is eventually found it may well prove to

Sieve tubes contain protoplasm which may circulate within each cell. It has been suggested that materials in this stream might pass through the sieve pores into adjacent cells. This would account for substances moving in either direction at the same time.
There is some evidence that such movement occurs, but not necessarily in the same tube.

Some workers have observed strands of protoplasm extending through many cells. These may play a part in the conduction of substances.

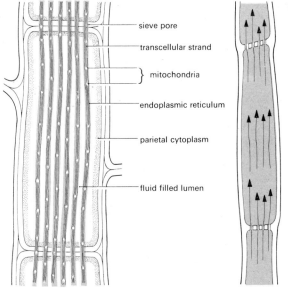

sieve pore

transcellular strand

} mitochondria

endoplasmic reticulum

parietal cytoplasm

fluid filled lumen

Mass flow theories propose that materials flow through phloem cells like water through a pipe.

Figure 39
Three hypotheses of transport in plants represented diagrammatically.
Diagram is based on Thaine, R. (1964)
'A protoplasmic-streaming theory of
phloem transport,' Journal of Experimental Botany *15*, p. 770.

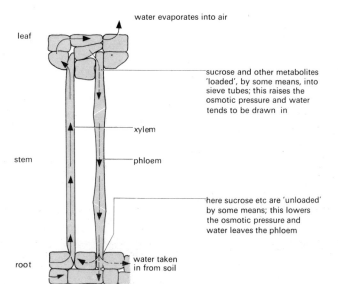

water evaporates into air

leaf

sucrose and other metabolites 'loaded', by some means, into sieve tubes; this raises the osmotic pressure and water tends to be drawn in

xylem

stem

phloem

here sucrose etc are 'unloaded' by some means; this lowers the osmotic pressure and water leaves the phloem

root

water taken in from soil

Figure 40
Hypothetical circulation in a plant.

be neither mass flow nor activated diffusion, but some method quite unforeseen at present.

Bibliography

Best, C. H. and Taylor, N. B. (1964) *The living body*. 4th edition. Chapman & Hall. (Human circulation from a medical point of view.)

Fogg, G. E. (1963) *The growth of plants*. Penguin. (Plant physiology.)

Freeman, W. H. (1970) Nuffield Advanced Biological Science Topic Review *Circulation*. Penguin. (Circulation in animals.)

Nuffield O-level Biology (1966) Text Year IV *Living things in action*. Longman/Penguin. (Fundamentals of blood systems of animals, including humans.)

Richardson, M. (1968) Studies in Biology, No. 11, *Translocation in plants*. Edward Arnold. (Plant physiology.)

Schmidt-Nielsen, K. (1960) *Animal physiology*. Prentice Hall. (Circulation in animals.)

Synopsis

1 Once materials have been taken in by an organism from its environment, there has to be a medium of transport so that they can be carried further. In most animals the medium is blood, and this is first examined as a means of carrying oxygen.

2 The red pigment of blood, haemoglobin, plays a vital part in oxygen transport. Its properties can be investigated experimentally; so also can its reactions with a number of other common gases.

3 The sap of plants and the blood of animals carry a great variety of substances besides gases. A quick test enables the transport of glucose to be investigated quantitatively.

Chapter 4

Transport media

Summary of practical work

section	topic
4.1	Examination of human blood
4.2	The carriage of oxygen
4.3	Haemoglobin reactions
4.4	Transport of materials other than gases in plants and animals

Organisms exchange materials with their environment through a variety of special systems; for instance, dissolved substances are conducted up and down the stems of plants as sap, a colourless liquid. The circulating fluid in animals, blood, has attracted man's attention since the dawn of history and probably long before, by its colour as much as for its biological importance. In some animals such as locusts the blood is colourless and hard to find, in a few it is green (some marine worms possess the pigment chlorocruorin) or blue (some molluscs contain haemocyanin), but in the great majority it is red, due to the presence of haemoglobin or a closely related compound.

The circulation of a fluid, whether it is water or blood, through an organism, provides a means of co-ordination as well as supply. The chemical activities of one part of the body are linked with those of all the other parts. A large animal is an 'organism' rather than a mere assemblage of tissues because it has a medium of transport circulating inside it.

Living cells and tissues have many chemical requirements, but one in particular, oxygen, is in continuous demand. In other words, if cells are deprived of oxygen for even a short time, they die. The continuous supply of oxygen to all tissues constitutes one of the major features of an animal's organization. Bearing this in mind, we should not be surprised to find that blood appears to be more adapted to carrying oxygen than any one of the other substances that it normally transports.

4.1 Examination of human blood

Blood is the easiest of the body tissues to examine under a microscope since it needs little preparation; even so, the observer needs to take care and think, otherwise what he sees may mislead him.

Procedure

When blood is removed from the body some of its physical and chemical properties change quite rapidly. For this reason it is desirable to use fresh samples. Stored blood contains additives to prevent clotting; refrigeration causes damage to cells. You can easily obtain small samples of fresh blood by pricking your finger.

1   Shake the hand vigorously (pointing it downwards) for a few seconds.
2   Select an area at the tip of a finger or thumb on the lower surface, swab it with sterile cottonwool soaked in ether or 70 per cent alcohol, and allow it to dry.
3   Jab the finger firmly with a sterile lancet; a good sized drop of blood should appear at once. If it does not, prick again more deeply. Do not continue trying to wring blood from an inadequate puncture or it will probably clot.
4   Never rest the sterile lancet on the bench or elsewhere, or let it be used by anyone else. Once it has been used, throw it away.
5   Let a drop of blood fall onto the centre of each of five clean slides and cover each drop with a cover-slip. (Bleeding from the thumb or finger will cease almost immediately and there is no need to wash it or cover it with plaster unless you intend to work with dirty or poisonous materials shortly afterwards.)
6   Put a drop of a different kind of liquid at one side of each of four of the cover-slips. When some mixing has occurred (figure 41), examine a region of unaffected blood and one of blood/liquid mixture. Use a microscope with H.P. magnification ($\times 400$).
    Suitable liquids which might be used include distilled water, sodium chloride solution (0·89 g dissolved and made up to 100 cm^3 with distilled water), household liquid detergent,

Figure 41
A simple experiment with blood.

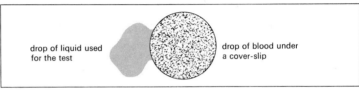

drop of liquid used
for the test

drop of blood under
a cover-slip

microscope slide

region where liquid and blood mingle

microscope slide

saturated salt solution, or strong sucrose solution.
Leave one slide of blood with no liquid added as a control.

7   Insert a micrometer graticule into the eyepiece of your microscope. Observe each slide in turn under H.P. and note the variety of sizes and shapes of the cells as soon as the preparations have been made, and again after 10–15 minutes.

8   To make a permanent preparation, put a drop of fresh blood near one end of a clean, grease-free slide. Take another clean slide and, holding it at an angle, *drag* the drop of blood across the slide (figure 42).
Allow the smear to dry by waving the slide about in the air but do not breathe on it. Examine it under H.P.

Questions

a   How many different kinds of corpuscle can be seen in untreated blood?

b   What is the size (diameter and thickness in microns) of the erythrocytes or red blood corpuscles?
Note that when you view an object with H.P. magnification you may see optifacts. These are due to the properties of light, not to imperfections of the microscope or preparation (figure 43).

c   How are the erythrocytes grouped together? Illustrate your answer with a simple, accurate drawing.

d   In what ways are the corpuscles affected by the various liquids used in stage (6)? Can you account for your observations? Seek information from textbooks if necessary (see Bibliography).

e   How does the blood of the dry smear differ in appearance from a drop of fresh blood?

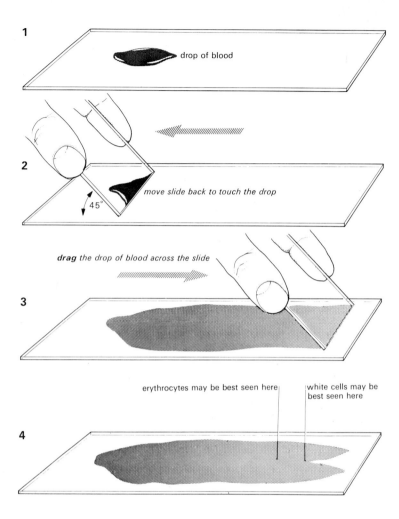

1

drop of blood

2

move slide back to touch the drop

45°

drag the drop of blood across the slide

3

erythrocytes may be best seen here   white cells may be
best seen here

4

Figure 42
Making a blood smear.

4.2 The carriage of oxygen

Blood performs many functions, but the previous chapters
have been concerned only with its ability to transport
oxygen. Before proceeding further we should first consider
the requirements of any transport system. To transport
something efficiently we need a medium which
a. moves rapidly
b. has a large capacity
c. is able to 'load' and 'unload' the transported material.

In the last chapter we saw that blood moves rapidly but its
capacity and loading properties are not easily detected.

Most gases dissolve in water to some extent, depending on
conditions of temperature and pressure. Though we warm
water when trying to dissolve solids such as sucrose, the

Figure 43
Optifacts and red corpuscles.

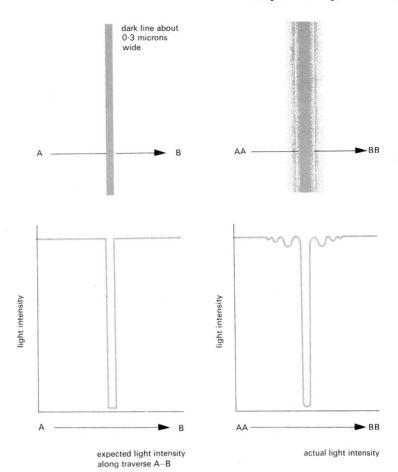

dark line about
0·3 microns
wide

expected light intensity
along traverse A–B

actual light intensity

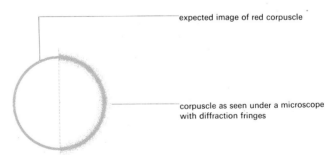

expected image of red corpuscle

corpuscle as seen under a microscope
with diffraction fringes

reverse is true for gases; the higher the temperature the *less* readily do they dissolve. Conversely the greater the pressure the more soluble they are. When we remove the cork of a lemonade bottle the pressure inside is reduced and carbon dioxide comes out of solution. Each gas has its characteristic tendency to dissolve. For example at 0° C and 760 mm pressure (1 atmosphere) 1 cm³ of water, exposed to pure nitrogen, will contain 0·024 cm³ of gas. The corresponding figure for 1 cm³ of water exposed to oxygen is 0·049 cm³. These amounts will not be taken up by the water if the two gases are present together at the same time. This is illustrated by the following example.

Suppose air consists of 79 per cent nitrogen and 21 per cent oxygen and some is well shaken with 1000 cm³ of water, at 0° C, 760 mm pressure. The nitrogen which dissolves will not be $1000 \times 0·024 = 24·0$ cm³ but 79 per cent of this amount, or 19·0 cm³.

Similarly the oxygen which dissolves will be:

$$1000 \times 0·049 \times \frac{21}{100} = 10·3 \text{ cm}^3$$

It is important to understand this principle because blood is exposed not to pure oxygen but to air which is a mixture of gases. If we depended on pure water as our medium of transport then we should require 1 dm³ for every 10·3 cm³ of oxygen at 0° C, and more than this at our higher body temperature. Table 1 in Chapter 1 gives figures for air-equilibrated water at higher temperatures. Using the results obtained with the spirometer it is quite easy to calculate the approximate volume of water which would be needed to remove the oxygen from the air inhaled by a man in one minute. Such a volume is enormous, and would be approximately 20–30 dm³ if it were moving at the same speed as blood. Alternatively, a smaller quantity would have to be pumped at a proportionally higher speed. From this we can make a reasonable guess that a given quantity of blood must be able to contain a much larger quantity of oxygen than the same volume of water.

We can get some idea of this capacity by displacing oxygen from blood by chemical means. We know of a number of reagents which do this; one of them is potassium ferricyanide. The procedure is essentially simple and involves mixing blood with ferricyanide solution and measuring how much the volume increases because this produces gaseous oxygen. The apparatus you need depends on the amount of blood available. If you have access to several cubic centimetres of blood, suitably treated to prevent clotting, then you should adopt Procedure A below. If you can only use a drop of blood taken from a finger then try Procedure B.

Procedure A

1   To measure small changes in gas volumes it is convenient to use a simple respirometer. One such device is described further in Chapter 6, when we consider the evolution of carbon dioxide in fermentation (page 112). Assemble the parts of a simple respirometer (figure 44).

2   Obtain between 1 and 3 cm³ of blood and measure its volume. Shake it well to allow air to dissolve in it. For every cubic centimetre of blood to be used, prepare 1·5 cm³ dilute ammonium hydroxide solution to absorb any carbon dioxide evolved, and 0·2 cm³ potassium ferricyanide solution.

3   Obtain a few drops of liquid detergent or a generous knife point of saponin, a substance with similar action, in order to release red pigment from the corpuscles and make the reaction with ferricyanide easier.

4   Modify the simple respirometer so that you can put a mixture of blood, hydroxide solution, and detergent in one vessel

Figure 44
A simple respirometer.

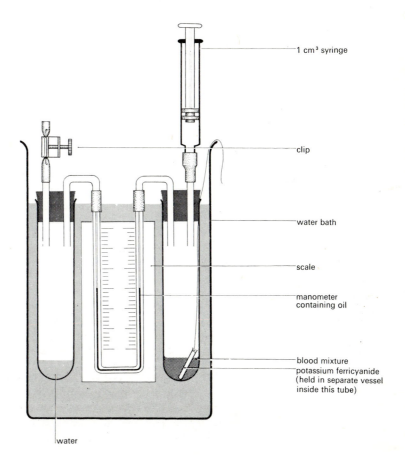

1 cm³ syringe

clip

water bath

scale

manometer containing oil

blood mixture potassium ferricyanide (held in separate vessel inside this tube)

water

so that it does not mix with the potassium ferricyanide.

5   Set up the apparatus in a waterbath at 30° C. When all movement of the manometer fluid has stopped, allow the ferricyanide solution to mix with the blood and observe the manometer fluid. When movement has ceased measure the difference in levels.

Procedure B

1   Take a piece of glass capillary tubing of about 1 mm bore and 40 cm in length. Heat the region between 5 and 10 cm from one end in a hot Bunsen flame until part of the tube bends over to form a U (figure 45).

2   When the tube is cool, attach a short length of transparent plastic (PVC) tubing to the short end and a 2 cm³ syringe to the other (figure 45). Obtain another piece of 1 mm bore capillary about 30 cm long.

3   Fill the U of the first tube with potassium ferricyanide (5 per cent) solution. Fix a 1 cm³ syringe to the straight capillary.

4   Obtain a large drop of blood from the thumb as described previously and allow it to fall onto a slide or watchglass. Draw it up into the straight capillary and measure the length; do this quickly.

5   Return the blood to the watchglass or slide and mix it with a trace of liquid detergent and a few drops of dilute ammonium hydroxide. The blood should be homogeneous; if it contains visible lumps or granules, reject it.

6   Draw a small drop of water into the straight capillary, then the blood mixture (see figure 45). Fix the straight capillary into the connecting tube. Remove the 1 cm³ syringe and immerse the whole U-tube into a gas jar or tall beaker of warm water at 30° C.

air bubble separation.

7   When there is no further movement of the meniscus (M) withdraw the piston of the 2 cm³ syringe quickly so that the blood mixture descends into the connection tubing, so mixing with the potassium ferricyanide solution. Stop when all the blood has passed out of the straight capillary but the water drop is still in it. Note the position of the meniscus (M), now and during the next five minutes. Record its final position.

Questions

a   From results obtained through procedures A or B, what was the percentage of gas displaced from the blood sample by potassium ferricyanide?

b   If we assume that the gas is oxygen, do your results confirm the hypothesis that blood can contain more oxygen per unit of volume than water?

c   How would you set about testing the composition of the gas displaced from a sample of blood?

With an adequate supply of blood it would not be difficult to separate the red corpuscles from the liquid (plasma) and

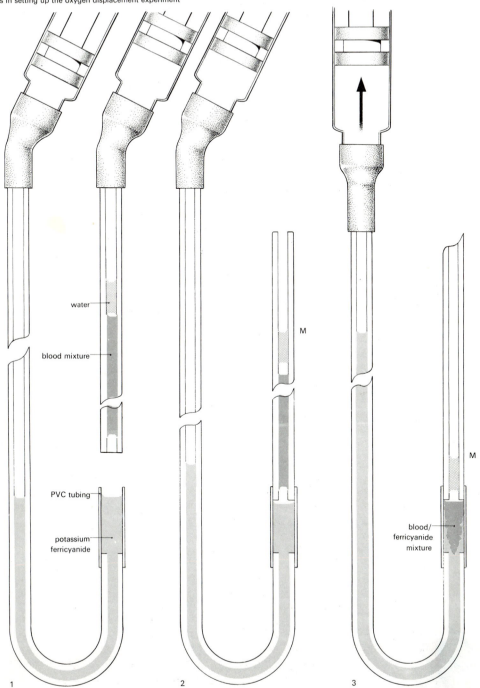

Figure 45
A device for measuring the oxygen in blood.

Three stages in setting up the oxygen displacement experiment

show that it is the material of the former (haemoglobin) which is associated with the great oxygen-holding property of blood. Some animals have haemoglobin free in the blood, not contained in corpuscles. By referring to textbooks (see Bibliography) find out:

d   Which groups of animals have 'free' haemoglobin and which have red corpuscles?

e   Is an advantage gained by animals which have their haemoglobin in corpuscles and, if so, what is it?

f   Are the erythrocytes of all animals shaped like human red cells? Is a biconcave disc a shape well adapted to the carriage of oxygen?

4.3 Haemoglobin reactions

It might be assumed from previous paragraphs that the main function of haemoglobin is to combine with oxygen alone. You can do a simple test to show whether this is true or not, if a supply of blood (10 cm³) is available.

Procedure

1   Pour about 2 cm³ of blood, to which citrate has already been added to prevent clotting, into a small test-tube. Swirl this gently to obtain a thin film of blood all over the inner surface of the tube.

2   Introduce a jet of oxygen, by means of a delivery tube from a cylinder, for a few seconds. Cork the tube and continue to rotate it to maintain a thin film of blood.

3   Repeat (1) and (2) with fresh blood, using a different gas each time. Possible gases are carbon dioxide, nitrogen, coal gas, air, hydrogen sulphide.

4   Compare the colour of the blood in each tube after about five minutes' exposure to the gas.

Questions

a   How many different colours of blood have you observed? Be careful to distinguish between different colours and varying shades or intensities of the same colour.

b   If we assume that each colour indicates a different compound of haemoglobin, how many such compounds have been formed?

c   Coal gas consists mainly of hydrogen and methane (about 80 per cent) and some carbon monoxide, ethylene, and nitrogen (20 per cent). Which of these gases do you think combines with haemoglobin? Account for your answer; you may need to refer to a textbook (see Bibliography). It is wise to find out, from your local Gas Board, the composition of gas supplied to the laboratory.

4.4 Transport of materials other than gases in plants and animals

If we look at a man-made transport system such as a railway

it is an easy matter to distinguish between the freight carried, the vehicles, and the rails. It is not so easy to see, in a plant's sap or an animal's blood, which materials are being carried and which constitute the medium of transport itself. It could be argued that all the materials in blood are being transported.

Whole blood consists of white cells, erythrocytes, platelets, and plasma. Plasma includes:

water	enzymes	*inorganic ions*
fats	proteins	*such as:*
phospholipids	urea	sodium
cholesterol	uric acid	calcium
glucose	creatine	potassium
amino acids	creatinine	magnesium
antibodies	ammonia	phosphate

If we want to know which of these are being transported we must find out which are being put into blood at one point in the system and removed at another. If we find there is a difference in concentration in any one of these substances at a point of entry into an organ and at an exit from it, this suggests that the substance is being transported and is not merely a permanent constituent of blood. To investigate this we need to make specific and quantitative tests for each plasma constituent. Though elaborate techniques are required for many of these components, we can identify glucose specifically using an enzyme/colour system prepared on a strip of absorbent cellulose.

Procedure

1   Obtain a large drop of blood from your thumb in the manner previously described and spread it over the coloured test area on the printed side of a Dextrostix reagent strip.
2   Wait exactly one minute.
3   Quickly wash the blood off the strip with a fine jet of cold water.
4   Compare the colour of the test area with the colour chart provided on the container. Do this *at once*. The blocks of colour represent concentrations of glucose in blood in mg per 100 cm^3 of blood, 40, 65, 90, 130, 150, 200, or over.
5   Cut off a shoot of a potted plant such as *Impatiens* and wait a few minutes for a drop of watery liquid to accumulate at a cut surface.
6   Test this drop as if it were a drop of blood (stages 1–4).
7   Put another drop of sap on a slide, cover with a cover-slip, and examine under H.P.
8   Consult a textbook which has details of the blood circulation of a mammal (see Bibliography) and find two large blood vessels which are likely to contain high and low concentrations of glucose respectively.

9   When a *freshly* killed mammal such as a rat is available, obtain one drop of blood from each of the blood vessels and test for glucose, using Dextrostix. Such an opportunity will arise when you are carrying out the work of the next chapter.

Questions

a   What is the concentration of glucose in blood from your thumb?
b   Is glucose present in plant sap; if so in what concentration?
c   Does sap contain material visible under the microscope? If so, describe it and try to interpret what you see.
d   From which vessels of the mammal did you sample blood? Was there a difference in glucose concentration? What can you infer from this result?

Bibliography

Allison, A. C. (1956) 'Human haemoglobin types'. *New Biology* **21**, 43–58. Penguin. (The carriage of oxygen.)
Best, C. H. and Taylor, N. B. (1964) *The living body*. 4th edition. Chapman & Hall. (Comprehensive introduction to blood from the medical standpoint.)
Freeman, W. H. (1970) Nuffield Advanced Biological Science Topic Review *Circulation*. Penguin. (Circulation in animals.)
Grove, A. J. and Newell, G. E. (1961) *Animal biology*. University Tutorial Press. (Circulation of a mammal.)
Schmidt-Nielsen, K. (1960) *Animal physiology*. Prentice Hall. (Brief outline of gas transport.)
Yapp, W. B. (1939) *An introduction to animal physiology*. Oxford University Press. (The action of haemoglobin and other respiratory pigments.)

Synopsis

1 Circulation in an animal is a dynamic process involving the movement of blood through a network of vessels. Complete understanding of the system demands the investigation of living animals, but much can be gained from the study of dead ones.

2 Dissection can reveal the route taken by blood through an animal body. The anterior part of a dogfish is a suitable subject for such a study because the relation between blood circulation and the gills can be established. This follows logically from Extension Work 1.

3 The course of the blood can be followed, in part, by dissection, but the smallest elements of the system require the use of a microscope and sectional material.

4 The way in which the heart of a dogfish functions can be deduced, to a large extent, from a critical examination of its anatomy.

Transport inside organisms: circulation in the dogfish

Summary of practical work

section *topic*

E2.1   Exposure of the heart

E2.2   Blood vessels joining the heart and gill system

E2.3   The drainage of blood from the gills

E2.4   Further study of the heart

E2.5   Passage of blood through the gills

In Extension Work I a method was described for investigating the gill system of a dogfish. That and the dissections described in Chapters 2, 3, and 5 involve relatively simple techniques and provide little indication of the skill required for a thorough anatomical investigation. In the following dissections, however, you will need precision and patience if you are to obtain useful information. Given time and an unlimited supply of fish it might be possible to achieve a good dissection by trial and error, but in normal working conditions some kind of guide is necessary. The following procedure is intended as a guide rather than a set of instructions to be slavishly obeyed. There are many alternative ways of dissecting, but always there must be a clear immediate objective at each stage as well as a longterm aim. As in investigation E1.1 you are advised to look at the questions before you begin and appreciate the reason for each instruction as you dissect.

The ultimate aim of the dissections that follow is to discover part of the course taken by oxygen from the gills of a dogfish to the rest of the body. Fish have blood rather like our own so it is a reasonable hypothesis that oxygen is carried in the blood as it is in human beings. If we can show that there are blood vessels leading to and from the gill system, a pump to move the blood, and valves to impose a direction of flow, this anatomical evidence will support the hypothesis that blood circulates through the gills in the living animal. We shall then be in a position to follow the course of oxygenated blood throughout the whole body.

E2.1 Exposure of the heart

Though it is possible to begin at the branchial arches (see E1.2) and follow the course of blood vessels, it is easier to start the dissection by exposing the heart and, later, establishing connections between it and the gill system.

Procedure

1   If you have not already done so, make incisions in the head of a dogfish to expose the pharynx, as shown in figures 21, 22, and 23 of investigation E1.1.

2   Make four incisions through the mucous membrane of the deflected part as indicated in figure 46. Peel back the strip of mucous membrane freed by these cuts; this can be done in one swift movement by grasping the front end of the strip between finger and thumb and ripping it off.

3   In the region of the fourth and fifth branchial pouches cut through the exposed tissue until you meet firm cartilage. This is the basibranchial cartilage. Refer to figure 46.

4   Scrape outwards from the mid-line to expose the whole extent of the basibranchial cartilage; do this quite boldly and quickly.

Figure 46
Diagram of branchial cartilages.

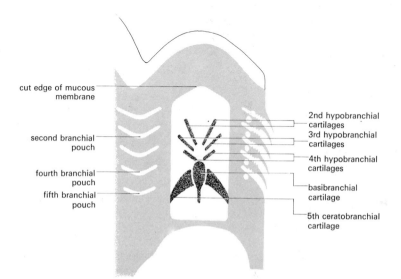

cut edge of mucous membrane

second branchial pouch

fourth branchial pouch

fifth branchial pouch

2nd hypobranchial cartilages

3rd hypobranchial cartilages

4th hypobranchial cartilages

basibranchial cartilage

5th ceratobranchial cartilage

5   Extend the exposed area by scraping forwards and laterally until you can see the second, third, and fourth hypobranchial cartilages and the fifth ceratobranchial of both sides. The scraped area should now resemble figure 46.

6   Carefully dissect the basibranchial and both the cerato-branchials; to avoid damaging the underlying heart you should hold the anterior end of the basibranchial with large forceps and lift it up as far as you can while cutting along its edges. To remove the ceratobranchials cut them and lift up the cut end as you dissect them.

7   Trim the cut edge of the pericardium taking particular care at the posterior margin as the sinus venosus closely adheres to the pericardium. (See figure 47.)
      You must be very careful not to trim away too much of the wall at the anterior end of the pericardial cavity, otherwise you will damage the fifth afferent branchial arteries which arise here from the ventral aorta. (See figure 47.)

8   You are now in a position to identify the four chambers of the heart. To display the ventricle you will have to deflect the atrium to one side and pin it in the deflected position. Locate the ductus Cuvierii of one side (the two ductus Cuvierii open into the sinus venosus laterally) and the posterior cardinal sinus; with a blunt seeker probe from the posterior cardinal sinus to establish the route by which blood enters the heart. Note the arrangement and size of the chambers and the nature of their walls.

Give reasons for each answer to the following.

Questions

a   Will blood pressure be high or low in the sinus venosus?
b   What causes blood to flow into the atrium from the sinus venosus?
c   What is the function of the atrium?
d   How is a uni-directional flow of blood ensured?
e   Which chamber forces blood out from the heart and drives it through the vascular system?
f   How is the chamber you selected in (e) adapted for its function?
g   Will the pressure of blood leaving the heart be high or low?
h   What purpose is served by the conus arteriosus?
i   How are its walls adapted for this function?
j   What prevents the wall of the pericardial cavity from moving?

E2.2 Blood vessels joining the heart and gill system

We can now see that the exposed heart is quite close to the gill bars which, as we know from E1.2, contain blood vessels. The next phase of the dissection is to see if there are connections between the heart and the gills.

Procedure

1   Remove the second left-hand hypobranchial cartilage (it will be on your left).
2   Clear the connective tissue along the mid-line until you expose the lateral border of the ventral aorta. This region has been chosen because the aorta gives off no branches opposite the second branchial pouch.
3   When the ventral aorta is clearly displayed, trace it forwards until you reach the right and left innominate arteries to which it leads. (See figure 47.)
4   Free the left innominate artery from connective tissue and trace it as far as you can. The pillar muscles (coracobranchials) are an obstacle to clearing.
5   Separate the first pillar muscle from connective tissue and from the innominate and when it is entirely free, cut through it as low down as possible and remove it. This should be done with scissors in a single operation. Try to leave a tidy stump.
6   Dissect the innominate out as far as its fork into the first and second afferent branchial arteries. Clear the bases of the first and second afferent arteries.
7   Remove the third hypobranchial cartilage; the third afferent branchial artery lies immediately beneath the cartilage.
8   Free the third afferent branchial artery from connective tissue.
9   Locate and free the second pillar muscle and remove it with a single scissor cut to leave a tidy stump, taking care to avoid damaging the second afferent artery.
10   Remove the fourth hypobranchial cartilage; again, the blood vessel is beneath it.

Figure 47
Diagram of the heart and ventral aorta.

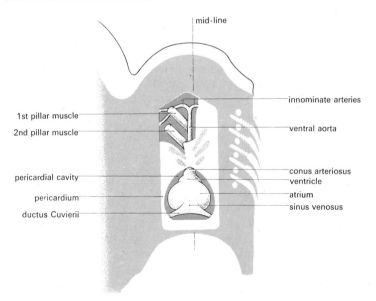

11   Free the fourth afferent branchial artery from connective tissue.

12   Remove the third and fourth pillar muscles; the fifth afferent branchial artery lies just posterior to the fourth pillar muscle. This is the most difficult of the blood vessels to dissect out neatly.

13   Now trace the third afferent branchial artery as far as possible. First, you will need to trace it to the beginning of the branchial bar along which its course lies. Once this is done it is not too difficult to dissect the artery along the arch provided you keep your scalpel close to the anterior border of the blood vessel and cut upwards all the time. By doing this you will free the artery from the third ceratobranchial and epibranchial cartilages which lie anterior to the artery (see figure 48).

(*Note:* If your first attempt to trace an afferent artery is unsuccessful, bear in mind where you ran into difficulties and try to trace one of the other arteries. It does not matter which one you finally succeed in tracing so long as you modify the next part of the investigation to fit in with your dissection.)

14   Having displayed one of the afferent branchial arteries, examine it carefully for tributaries and note the general direction of such branch vessels as you see.

15   Assess the relative thickness of the walls of the ventral aorta, the afferent branchial arteries, and the posterior cardinal sinus. To do this you will have to remove pieces of blood vessel. If you found tributaries of the afferent branchial artery in the gill bar region, note the possible course for these vessels relative to the gill lamellae and the septum.

Figure 48
Diagram of the afferent branchial arteries.

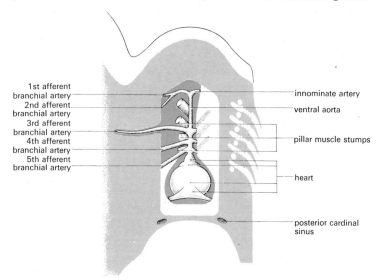

1st afferent branchial artery

2nd afferent branchial artery

3rd afferent branchial artery

4th afferent branchial artery

5th afferent branchial artery

innominate artery

ventral aorta

pillar muscle stumps

heart

posterior cardinal sinus

Questions
You can answer these in part by reasonable hypotheses based on examination of your dissection. To give complete answers or confirm these hypotheses you will need to refer to text-books.

a   What do you consider to be the function of the pillar muscles?

b   What purpose is served by the various cartilages of the branchial skeleton?

c   How are the pillar muscles related to the branchial skeleton and the pectoral girdle?

d   What correlation is there between thickness of wall and likely blood pressure?

e   What sort of blood, oxygenated or deoxygenated, do you think flows in the vessels you have dissected?

f   For what purpose is blood being pumped to the branchial bars?

g   Make an illustrated record of the parts of your dissection which show the probable course of blood from the sinus venosus to any one branchial bar.

E2.3 The drainage of blood from the gills

The dissection so far shows the way blood probably travels to the gills. If we are to understand how a living dogfish conveys oxygen, absorbed from sea water, into its body tissues, we must find the course of blood *away* from the gills. We should expect to find fairly conspicuous vessels draining these important organs and carrying oxygenated blood for the whole body. As to the position of such vessels, two possi-

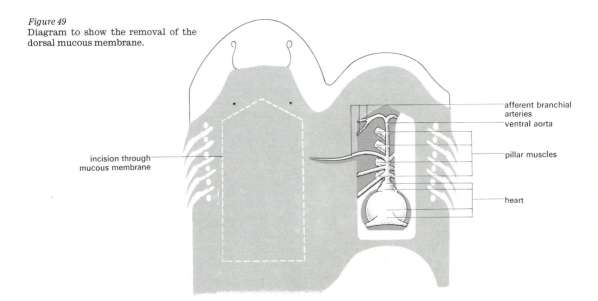

Figure 49
Diagram to show the removal of the dorsal mucous membrane.

incision through mucous membrane

afferent branchial arteries
ventral aorta
pillar muscles
heart

bilities exist. One is that they double back from the gills and are situated in the ventral portion of the fish with the heart. You will not have observed any while dissecting this region, apart from the afferent branchial arteries. We must therefore explore the other possibility, that they are in the dorsal portion of the fish which lies above the pharynx.

Procedure

1 Make four incisions through the mucous membrane of the *dorsal* surface of the pharynx (figure 49) and rip off the mucous membrane as for the ventral surface.

2 Locate the four pharyngobranchial cartilages on the animal's left side by scraping the exposed area until they appear; when you have found them look for the epibranchial arteries. The first artery is tight against the anterior surface of the first pharyngobranchial cartilage, and the remaining three lie clear from their related cartilages (about 3 mm anterior to them).

3 Remove the pharyngobranchial cartilages. The first pharyngobranchial of the left side may join with the corresponding cartilage of the other side in which case you should cut a small part of the right cartilage away.

4 Locate and clean the dorsal aorta. *Either* find the route of blood going to the right pectoral fin, *or* trace the aorta to the tail if this has not been removed previously.

5 Trace the epibranchial arteries from the dorsal aorta to the branchial bars. Refer to figure 50.

In order to follow the course of the arteries into the branchial bars it is necessary to cut away the mucous membrane, cartilages, and muscle tissue of the bar. This should now be attempted on the second branchial bar.

Figure 50
Diagram of the dorsal aorta and epibranchial arteries.

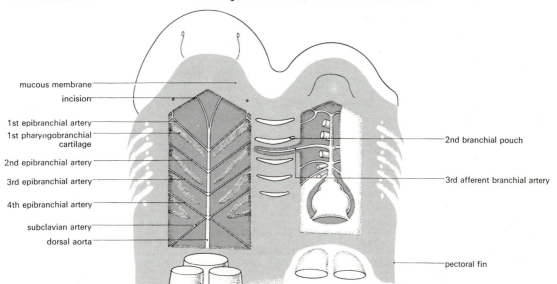

mucous membrane
incision
1st epibranchial artery
1st pharyngobranchial cartilage
2nd epibranchial artery
3rd epibranchial artery
4th epibranchial artery
subclavian artery
dorsal aorta

2nd branchial pouch
3rd afferent branchial artery
pectoral fin

6   Identify the ceratobranchial and epibranchial cartilages of the second bar and carefully remove them.

7   Remove any muscle tissue that obscures the view. Peel the anterior part of the mucous membrane of the second bar forward. (See figure 51.)

8   Lift up the flap of mucous membrane that you peeled forward and you should be able to see the efferent branchial artery. When you have identified this (it usually contains enough blood to make this possible but occasionally it is bloodless), trim the mucous membrane down to the blood vessels.

9   Try to find the whole of the efferent loop from the second epibranchial artery; this goes right round the second branchial pouch, so you will have to dissect the posterior surface of the first bar to find the rest of the loop. This is done by making a median incision through the mucous membrane of the first bar and pulling the posterior half of the mucous membrane back until you can see the efferent branchial artery. When you have found it trim the mucous membrane with scissors down to the level of the blood vessel. (See figure 51.) The dissection is now completed and it should look something like figure 52.

Questions

a   Which have the thicker walls, the epibranchial arteries or the afferent branchials? (To find the answer you will have to remove and examine pieces of each vessel.)

Figure 51
Diagram to show the technique for exposing an efferent loop.

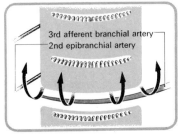

Stage 1
3rd afferent branchial artery displayed

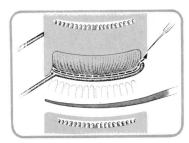

Stage 2
posterior part of the 2nd efferent loop

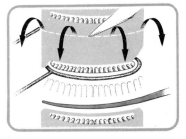

Stage 3
Incision down the middle of the 1st bar

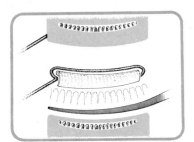

Stage 4
anterior part of the 2nd efferent loop

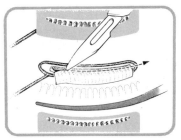

Stage 5
trimming the 2nd efferent loop

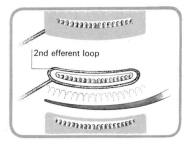

Stage 6
the finished dissection

b   How do you account for any difference in thickness?.
c   What sort of blood, oxygenated or deoxygenated, do you think the epibranchial arteries contain?
d   By what route is this blood distributed to the rest of the body?
e   How do you think the blood pressure in the dorsal aorta compares with that in the ventral aorta? Account for the differences you suggest.
f   How does blood from the heart reach *either* the right pectoral fin *or* the muscles of the tail? Make a *diagram* of your dissection with arrows to indicate the direction of blood flow.

Figure 52
Diagram of the finished dissection.

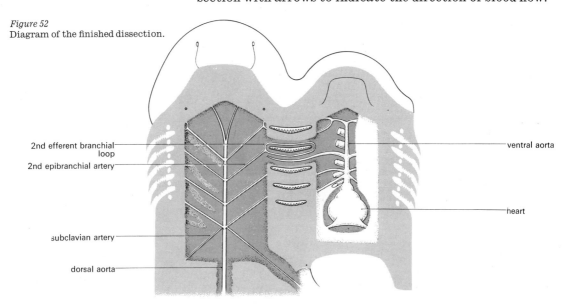

2nd efferent branchial loop

2nd epibranchial artery

subclavian artery

dorsal aorta

ventral aorta

heart

E2.4 Further study of the heart

We may now investigate two problems: how backflow of blood into the heart from the ventral aorta is prevented and how blood gets from the afferent arteries to the efferent. The first problem can be solved by removing and dissecting the heart; to solve the second you will need to examine sections of the branchial arch under a microscope.

Procedure

1   Cut through the fourth and fifth afferent arteries at their bases and through the ventral aorta anterior to the fifth afferent branchial. Dissect the ventral aorta and the conus arteriosus away from connective tissue.
2   The sinus venosus is attached to the pericardium and to free it you must find its borders and then cut this chamber of the heart away from connective tissue. Cut the Cuvierian ducts as far out as possible and remove the heart from the pericardial cavity.

3   Pin the heart to the dissecting board with the sinus venosus uppermost. Cut a window in the dorsal wall of the sinus venosus (i.e. the surface that is nearest to you) but be very careful to cut the dorsal wall only.

4   Observe the valves between the sinus venosus and the atrium. Note the number and position of the valve flaps.

5   Open the atrium and search for the valve between the atrium and the ventricle. Note the strands of muscle tissue in the atrial wall.

6   Turn the heart over and make a median incision through the ventricle, taking this cut up through the conus arteriosus and the ventral aorta. Pin out the conus arteriosus and ventral aorta, stretching them as much as possible. Examine the valves in the conus.

Questions

a   *1.* Is the sino-atrial valve symmetrically placed?
2. What prevents the sino-atrial valve from opening when the atrium contracts?
3. How many flaps are there in the sino-atrial valve?
4. How are the flaps arranged?

b   *1.* Is the atrio-ventricular valve symmetrically arranged?
2. What prevents the atrio-ventricular valve from opening when the ventricle contracts?
3. How many flaps are there in the atrio-ventricular valve?
4. How are the flaps arranged?

c   *1.* How many rows of valves are there in the conus arteriosus?
2. How many valves are there in each row?
3. Are the valves all the same?
4. What prevents these valves from opening when blood pressure in the ventral aorta exceeds ventricular pressure?

d   What is your theory of the action of the heart? Illustrate your answer with sketches or diagrams of the dissected heart. Does your theory of the action of the heart fit in with the direction of flow you indicated in the diagram of the arterial system?

E2.5 Passage of blood through the gills

Dissection reveals that blood vessels from the heart penetrate the gill system and that others join the gills to the remaining organs of the body. But dissection alone cannot show any connection between these two sets of vessels, the afferent and efferent arteries (or afferent branchial and epibranchial arteries). To see such a connection you will require a microscope and microtome sections of prepared gill tissue. If you have worked through investigation E1.2 you will already be familiar with the distribution of tissues in a section of a branchial bar. In the present context, we shall concentrate on the arrangement of minute blood vessels within the bar and the lamellae. You now have the difficult task of trying to understand a complex, three-dimensional structure by means of a two-dimensional slice.

Procedure

1 Refer to investigation E1.2 for information about the general structure of a branchial (gill) bar. Examine a prepared slide of a transverse section of a branchial bar with a hand lens (×10). Try to visualize the plane through which the section was cut; a model (see E1.3) is a useful aid.

2 Using a microscope search the central region of the section for tributaries of the afferent branchial artery which can be identified by its position, tubular nature, and contents. Inside it will be red blood cells, and in the dogfish these are large, ovoid, nucleated bodies.

3 Follow the tributaries of the afferent artery to the base of the primary gill lamellae. At this point you should have little difficulty in finding capillaries arising from the afferent tributary. There is a small dilation at the base of each lamella representing the smallest branch of the afferent artery and it is from this vessel that the capillaries arise.

4 Follow the capillaries through the secondary lamellae.

5 Examine the distal ends (i.e. the part furthest away from the centre of the section) of the secondary lamellae.

6 Make a sketch of a section of a branchial bar showing the details you have observed of the circulation in the secondary lamellae and how this bridges the gap between afferent and efferent branchial arteries. On this sketch indicate the direction of flow of the blood in the capillaries of the secondary lamellae and the direction of flow of water over the surface of the lamellae.

Questions

a How wide are the capillaries of the lamellae compared with the size of the red blood corpuscles?

b Using the diameter of a red blood corpuscle as a unit, compare the following: the thickness of a capillary wall, that of the outer layer (epidermis) of a secondary lamella, and that of tissue which lies between them.

c If the gill lamellae are responsible for the exchange of substances between sea water and the blood, what are these substances and how are the lamellae suited for such exchange?

d In which direction do you think the blood might flow in the capillaries of a secondary lamella compared with the water outside? Do you think that the exchange of substances between the sea water outside and the blood in the capillaries would be facilitated by flow in opposite directions or by flow in the same direction? Or do you think it would be the same in either case? Give the reasons for your answer.

e Can you suggest any reasons to account for the fact that dogfish die when taken out of the sea, in spite of the much higher concentration of oxygen in air compared with that in sea water?

f   Summarize the chief differences between the circulation of blood through the gills and body of a dogfish, and through the lungs and body of a mammal. Can the differences be related to the differing sources of oxygen, sea water and air?

Bibliography

Freeman, W. H. (1970) Nuffield Advanced Biological Science Topic Review *Circulation*. Penguin.

Grove, A. J., and Newell, G. E. (1966) *Animal biology*. 7th edition. University Tutorial Press. (Blood circulation of the dogfish compared with other vertebrates.)

Hughes, G. M. (1963) *Comparative physiology of vertebrate respiration*. Heinemann.

Kylstra, J. A. (1968) 'Experiments in water-breathing.' *Scientific American* **219**, 2, 66.

Wells, T. A. G. (1961) *Three vertebrates*. 2nd edition. Heinemann. (Comparative account of structure and function.)

Synopsis

1 A fundamental activity of organisms is feeding. Before it is absorbed, food is digested. Digestion is seen as a process occurring in micro-organisms and even in certain plant organs, and not merely restricted to animals with alimentary canals.

2 The entry of digested food through a boundary membrane can be studied by analogy with an artificial model.

3 An examination of the gross and fine structure of the mammalian gut enables its structure to be related to function.

Chapter 5

Digestion and absorption

Summary of practical work

5.1 Digestion by micro-organisms and tissues

We have seen that organisms can change the composition of their environments, for instance, by altering the composition of the atmosphere around them. As living creatures grow they must also obtain raw materials from the environment. Energy, too, must be transferred from the environment in one form or another, for the metabolic process of life. Whereas green plants take in only simple chemicals and use light energy for their metabolism, animals and many micro-organisms rely on complex chemicals which they take in as food. This type of feeding is termed *heterotrophic*. Creatures which, by contrast, need only simple chemicals are known as *autotrophic* feeders.

This chapter is concerned almost exclusively with heterotrophic organisms. The food they consume is seldom available in the environment in a form exactly suited to their metabolic requirements.

Starch is an example of a complex chemical commonly used as food. If we put micro-organisms in an environment rich in starch we can easily investigate what immediately happens to it. We can use the same technique, although it is not applicable to a large animal as a whole, to investigate the effect of animals' organs and tissues on food.

One way of making an artificial environment suitable for life is to mix a starch suspension with agar jelly. The surface is open to the air and the jelly contains water; these are two factors essential to life. To exclude all unwanted organisms such as bacteria we should have to use strict sterile culture methods, but we can obtain useful results without doing so. The following methods do not involve strict sterility.

Procedure

1   You will need six Petri dishes (with lids) each containing a suspension of starch in agar.
2   Place organisms and tissues on the agar surfaces or in wells cut with a cork borer (figure 53). Do not put more than four items on any one agar plate. Some possibilities are:
A drop of yeast suspension.
A small portion of the gut of a locust (see 2.2 for method of dissection).
Half a germinating grain of barley.
A drop of human saliva.
A drop of distilled, sterile water (control).
Store any plates you do not use in a refrigerator.
3   Label each dish carefully so that you can keep a record of the contents, then place the plates in an incubator for 48 hours at 25° C.
4   Remove the plates from the incubator, and take off the lids and pour a solution of iodine dissolved in potassium iodide

Figure 53
Preparing a starch/agar plate.

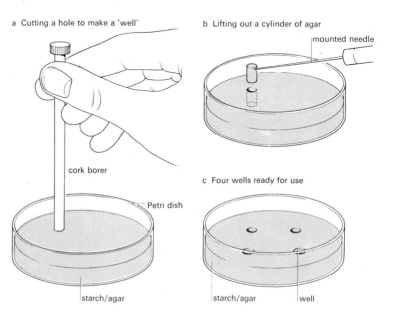

a Cutting a hole to make a 'well'

b Lifting out a cylinder of agar

mounted needle

cork borer

Petri dish

c Four wells ready for use

starch/agar

starch/agar   well

liberally all over the surface of the agar in each dish. Wait 15–30 seconds and then pour it away.

5   Examine the plates carefully, holding each in turn in front of a lamp, and record patches and zones of different colour. Place the used plates in a refrigerator for future reference.

Questions

a   Unchanged starch forms a deep blue colour with iodine solution. Which organisms or other items are associated with colourless zones?

b   Are such zones immediately touching the items themselves or do they extend beyond them?

c   What can you say about the chemical composition of the non-blue zones of agar? What further tests could you perform on these plates?

d   What other colours have you observed besides blue? How do you account for them?

e   In the light of your answers to (a), (b), and (c), relate this investigation to the process of digestion. If animal and plant tissue and micro-organisms can affect starch in the same way, is there any significance in this similarity?

5.2 Digestive organs: a model gut

Simple organisms can bring about chemical changes in any potential food that may lie around them. This does not necessarily imply that they use all the material for feeding but the nature of the chemical changes (digestion) at least suggests that the potential food may not be acceptable in its original form. In the case of starch we might ask how it is changed and why the digested product is more acceptable to organisms than starch itself?

Small organisms directly affect a starch environment; larger ones do not. It is only certain of their parts or organs which do this (figure 54).

We cannot understand the significance of the digestive process without reference to the skins and membranes which separate organisms from their environments. In the next investigation we shall use as a model an artificial membrane which is similar in some ways to the wall of the insect and mammal intestine. By using starch as before, we can consider the following procedure as an extension of the previous investigation.

Procedure

1   Dissect a freshly killed adult locust as directed (2.2). Examine the gut and then carefully remove it (figure 55).

2   Cut the gut into small pieces with scissors, put it in a watch-glass or cavity block, and cover it with a little starch suspension (1 per cent).

Figure 54
Diagrammatic comparison of the
action of a small and a larger organism
on food.

A micro-organism having no gut or special
feeding organs lies on a substrate of food

thin membrane

substrate containing food

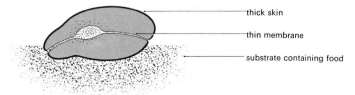

A larger organism ingesting food through a mouth
and digesting it in a simple alimentary tract (gut)

thick skin

thin membrane

substrate containing food

3   Cut the tissue using mounted needles or pointed scalpels.
4   Tie a knot in the free end of a roll of Visking (cellulose) tubing and cut 15 cm from the knot.
5   Pour the cut-up tissue into the tube and add more starch suspension until the total volume is about 10 cm³. Mix the tissue and suspension well by squeezing the tube between finger and thumb.
6   Lower the Visking tubing into a large test-tube (2·5 × 15 cm) containing about 10 cm³ of distilled water. Use a bulldog clip or paper clip to prevent the open end from falling in (figure 56).
7   Set up a similar Visking tube containing only 10 cm³ of starch suspension placed in a test-tube of water.
8   Incubate both tubes by placing them in a water bath at 25–30° C for 25–30 minutes.

Figure 55
A dissected locust showing the
alimentary canal.
After figure 27, Thomas, J. G. (1963)
Dissection of the locust, *Witherby*.

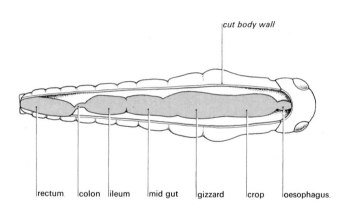

cut body wall

rectum   colon   ileum   mid gut   gizzard   crop   oesophagus

9 Remove a drop of fluid from the gut/starch mixture and add one drop of iodine solution on a white tile. Take a further 1–2 cm³ of the gut/starch mixture and add it to an equal volume of Benedict's solution in a test-tube. Heat the mixture until it boils. Record any changes of colour resulting from these tests.

10 Perform the same tests on:
The water surrounding the tube containing gut/starch.
The contents of the Visking tube.
The water surrounding the Visking tube.
Wash your pipette carefully after taking each sample to prevent accidental mixing of the materials being tested. Record your observations.

Questions

a Iodine forms a blue compound with starch; Benedict's solution forms an orange precipitate when heated with reducing sugars. In which samples did you find starch and in which did you find reducing sugars?

b From your observations alone, is there any evidence that gut tissue changes starch to reducing sugar? If so, explain it clearly.

c Can you add to this evidence by inspecting the starch/agar plates in 5.1?

d Consider the answers to question (a). What do these observations and the properties of cellulose tubing suggest to you about the problem of digestion for a living locust feeding on starch-filled leaves?

Figure 56
Using Visking tubing as a model gut.

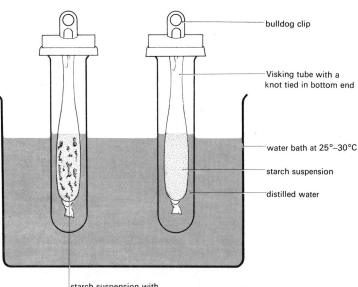

bulldog clip

Visking tube with a knot tied in bottom end

water bath at 25°–30°C

starch suspension

distilled water

starch suspension with fragments of macerated locust gut

5.3 The double function of the alimentary canal

So far, we have considered the locust gut as if it were a single, simple organ. But when we dissect and remove it and study figure 55 we see clearly that this is not so. Animals are usually covered by impermeable skins, and just as tracheal systems and lungs provide special surfaces for gaseous exchange, so alimentary canals have special surfaces concerned with feeding. Oxygen in the air is acceptable as such to organisms, but food has to be changed before it can be used. An alimentary canal, therefore, must have two functions, digestion and absorption. You should bear this in mind in the following investigation where we shall attempt to relate the structure and activity of the mammalian gut to its double function.

For the sake of simplicity, we have investigated only starch in the two preceding investigations. Natural food consists of many different chemicals, each of which has to be digested, and this is just as true for locusts and cattle eating grass as for humans who require a mixed diet. Complex alimentary canals are associated with foods which are difficult to digest. The gut of a ruminant, such as a cow, is more complicated than that of an omnivore, such as a rat. When we examine alimentary canals externally it is not possible to tell which parts are responsible for the digestion of protein, starch, cellulose, or fat; we have to perform chemical tests to discover this. As the contents of the gut pass through it from one end to the other, they mix, making it difficult to trace the chemical specialization of the various regions. Therefore, in the following examination of the alimentary system of a mammal, we shall restrict questions to a few that we can reasonably answer from a single, brief dissection. If we carry this out directly the animal has been killed and surround the gut by a medium similar in composition and temperature to that inside the living animal, then the alimentary tract is likely to continue moving for a short time in characteristic fashion, even though the animal is dead. This will provide more information than dissecting a preserved animal or one that has been dead for some time.

Procedure

Any small mammal is suitable; mice are often the easiest to obtain and they respond well to the treatment outlined below. Rats are best for stages (8)–(11).

1 Heat a quantity of Ringer's solution (at least 2 dm³) to a temperature of 36–38° C, and place it in beakers or jars in a thermostatically controlled waterbath set at 37° C.

2 As soon as the mammal has been killed, place it on its back in a wax-bottomed dish and rapidly fix it in position with pins.

3 Immediately cover the animal with warm Ringer's solution.

4 Working as swiftly as possible, open the abdomen by pulling

up a fold of skin in the centre with forceps, and then proceed
as in 2.4 (stage 2).

5   As soon as you have pinned back the flaps of skin and muscle
out of the way, stop dissecting and observe carefully the contents of the abdomen. Look for movement of any kind, using
a binocular microscope.

6   Check the temperature and replace the Ringer's solution if
necessary.

7   If you see movements, record your observations carefully,
and then proceed to identify the main parts of the alimentary
canal. These should be familiar from the work you did before
in 2.4. (See also figure 57.) This exploration may start more
intestinal movements, in which case stop and watch them.
There are at least three distinct patterns of movement known
to occur in the gut.

Figure 57
Diagrams of the alimentary canal of
rat and mouse.
After Rowett, H. G. Q. (1952) Dissection
guides III: The rat with notes on the
mouse, *2nd edition, John Murray.*

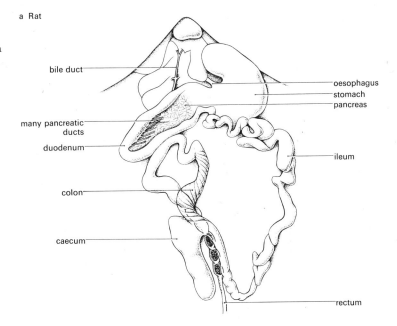

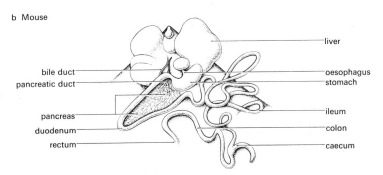

8   Remove the stomach, duodenum, ileum, caecum, colon, and rectum separately. Open them and wash out the contents of each part with warm Ringer's solution into separate labelled specimen tubes.

9   Return these parts of the gut to warm Ringer's solution and cut out small sections from the walls of stomach, duodenum, and ileum.

10 Examine each of these in turn, mounted in a drop of Ringer's solution under a binocular microscope, to discover the form of the internal surface.

11 Record your observations, as far as possible, by simple sketches.

12 Add a little distilled water to the contents of the specimen tubes, shake well, and leave to settle.

13 At a convenient time take a minute quantity of the clearer liquid above the sediment and place it 2 cm from the end of a prepared thin layer plate (figure 58). Use plates for the analysis of the contents of stomach, duodenum, and ileum. When they are dry, put the plates in a jar as shown, containing a mixture of n-butanol (40 parts), glacial acetic acid (10 parts), and water (15 parts), to a depth of 1 cm. Put a lid on the jar.

14 After one and a half hours take the plates to a fume cupboard, dry them, and spray thinly with ninhydrin solution.

15 Heat the sprayed plates to about 100° C in an oven.

16 Examine each plate carefully for purple and other coloured spots.

Questions

a   Which parts of the alimentary canal appeared to move; how do you account for the movement?

b   It is commonly known that food passes along the alimentary tract from mouth to anus. Do the gut movements appear to move food in this direction? If not, what did the movements appear to do to the food inside the gut?

c   Does the inside surface of the alimentary canal appear to be adapted to a digestive or absorptive function, or both? Apply the question to each part of the gut you examined, and state the reasons for your conclusions.

d   The solution rising up thin layer plates separates the various products of digestion of proteins; ninhydrin forms coloured compounds with amino acids. From your observations is there any evidence that protein has been digested?

e   Which part of the gut is chiefly responsible for absorbing digested food? (Confirm your answer by reference to a textbook.) It would be reasonable to assume that absorption occurs in the last part through which food passes in its journey along the alimentary canal. How do you account for the fact that this is not so?

Figure 58
Producing a thin layer
chromatogram.

Preparing a thin layer chromatogram

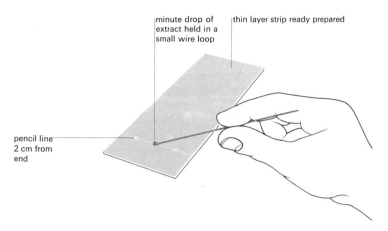

minute drop of
extract held in a
small wire loop

thin layer strip ready prepared

pencil line
2 cm from
end

Running a chromatogram

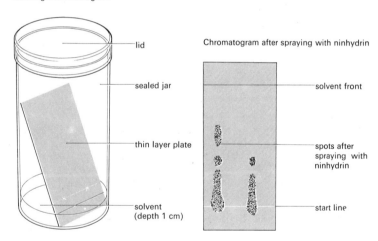

lid

sealed jar

thin layer plate

solvent
(depth 1 cm)

Chromatogram after spraying with ninhydrin

solvent front

spots after
spraying with
ninhydrin

start line

5.4 Fine structure of the intestinal wall

Pieces of locust gut bring about changes in starch (5.1 and
5.2); evidently something in the tissue digests food. The use
of starch/agar plates suggests that digestion is due to a substance or substances which diffuse out from the gut tissue.
It is common knowledge that alimentary canals secrete
digestive juices but it is not possible to see this actually happening when we look at a piece of intestine or stomach wall.
In the following investigation we examine a thin, stained
section of intestine wall in an attempt to relate the functions of secretion, absorption, and movement to structures
that we can see under the microscope. We can also obtain
help from photomicrographs and drawings of other tissues.
For example, the cells of glands have a characteristic shape

and size, and the presence of these in the gut wall could suggest a secretory function. Similarly, by examining sections of smooth (involuntary) muscle, we shall be able to identify such tissues if they are present in the intestine as they were in sections of blood vessels (see 3.5).

When studying mammal lungs we were concerned with the path of oxygen molecules from the air spaces into the blood. Blood contains glucose which is a digestive product of carbohydrates such as starch. As with lung tissue, we must look critically at the intestine to see by what path molecules of glucose can pass from digested food into the blood system. Remember that a section is two-dimensional but has to be interpreted as part of an original three-dimensional structure. It will be helpful if you examine pieces of gut from the dissected mammal first (2.5).

Procedure

1   Use a prepared, transverse section of the ileum of a rat or mouse. Note the staining technique employed, as indicated on the label.
2   Examine the section with the naked eye and with a hand lens ($\times 10$) or L.P. microscope. Compare the appearance with sketches made of the samples of gut from the dissected animal.
3   Look at the section under higher power magnification ($\times 100$, $\times 400$).

Questions

a   Describe the general form of the internal surface of the section. Does your description agree with that of the piece of ileum cut from the dissected animal?
b   Where are blood vessels situated in the wall of the intestine? Refer to figure 59, which is a photomicrograph of an injected specimen. By what route can digestive products pass into the blood stream of a living animal?
c   Do you consider the ileum to be an organ well adapted for the absorption of digested food? Give your reasons.
d   If your section is stained by P.A.S. (periodic acid, Schiff method), mucus and the cells producing it will show up clearly, stained deep pink. These are called goblet cells on account of their shape. Comment on their frequency and distribution; what part do you think mucus plays in the action of the alimentary canal? Can you identify other secretory cells? If so, record their position with the aid of a simple, outline sketch.
e   On the same or another sketch record the position of muscle tissue. If this contracts in the living animal, what movement of the intestine is brought about?
f   Look at figure 60, an electron micrograph of a minute portion of the inner surface of the ileum. There are about, 200 000 000 finger-like projections per square millimetre of surface.

Figure 59
Photomicrograph of ileum of cat with blood vessels injected (×110).
From Freeman, W. H. and Bracegirdle, B. (1966) An atlas of histology, Heinemann.

Comment on the possible significance of these structures with reference to question (c).

The investigations outlined in this chapter suggest that food is digested by organisms to a form which can pass either

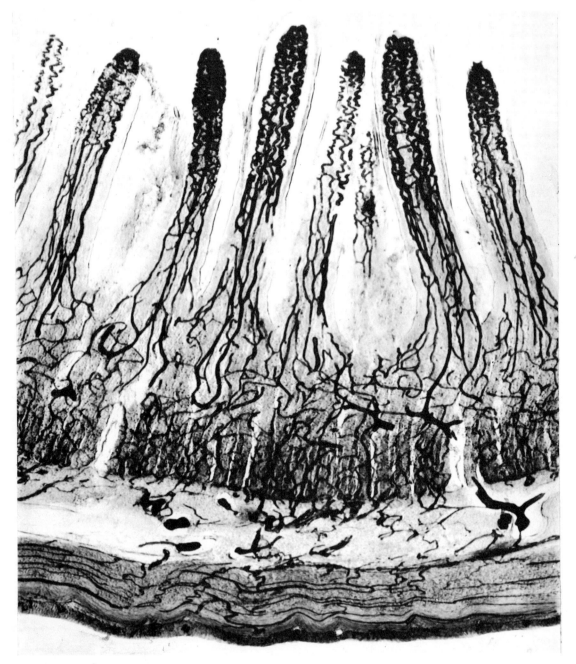

through all or through some specialized parts of the body surface, just as oxygen passes through lungs into the blood. There is strong evidence that the passage of digested food into the blood system is by no means as simple as it appears at first sight. Such evidence is derived from experiments outside the scope of this course but it must, nevertheless, be borne in mind when attempting to make hypotheses about the more detailed functions of the alimentary canal.

Bibliography

Best, C. H. and Taylor, N. B. (1964) *The living body*. 4th edition. Chapman & Hall. (Digestion applied to human beings.)
Clegg, A. G. and Clegg, P. C. (1962) *Biology of the mammal*. Heinemann. (General mammal digestion.)
Freeman, W. H. and Bracegirdle, B. (1966) *An atlas of histology*. Heinemann. (Helpful for interpreting prepared sections.)
Morton, J. (1968) *Guts*. Arnold. (A short and lively introduction to the subject.)
Schmidt-Nielsen, K. (1960) *Animal physiology*. Prentice-Hall. (Digestion.)

Figure 60
Electron microgrraph of the surface of the ileum of a rat (×14 900).
Photo, Professor D. Lacy.

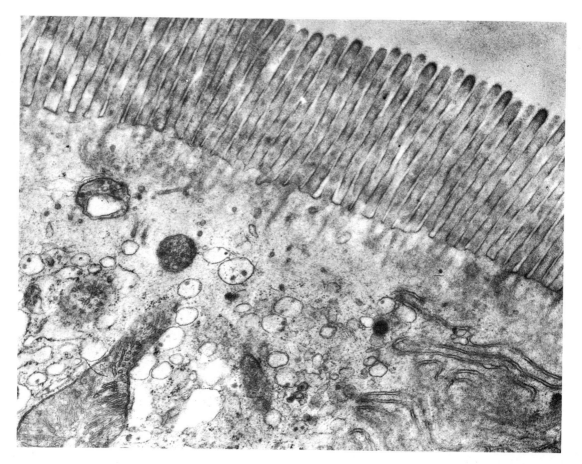

Synopsis

1 The enzyme responsible for the digestion of starch can be extracted from germinating cereal grain.

2 Using this extract, the course of starch digestion is followed quantitatively and some of the properties of the enzyme are investigated.

3 In living organisms enzyme-controlled reactions usually occur in systems rather than as single, independent reactions.

4 The role of an enzyme in the synthesis of starch can be studied. The enzyme is extracted from potatoes. The technique of study is that used with starch digestion.

Chapter 6

Enzymes and organisms

Summary of practical work

section	topic
6.1	Enzyme extraction
6.2	The course of an enzyme-controlled reaction
6.3	Enzyme activity in whole organisms
6.4	An enzyme-controlled synthesis

6.1 Enzyme extraction

All organisms, from the simplest microscopic bacteria to the most complex animals and plants, have a truly vast number of chemical reactions taking place in their bodies. Collectively these reactions constitute the metabolism of an organism. It is not feasible for us to list all the known reactions which constitute metabolic activity and try to understand all the chemical interactions within a single living cell. It is more profitable to study one reaction carefully, particularly as we now know that almost all metabolic reactions have one feature in common: they are controlled by enzymes.

In the work of the last chapter (see 5.1) it was shown that a variety of tissues can digest starch. Moreover, the plate technique showed that starch digestion occurs in a wide area around such tissues. This indicates that the tissues produce a substance capable of diffusing from the living cells that produce it. The substance is called an enzyme. The name literally means 'in yeast'; the first enzymes discovered were obtained from yeast.

In the nineteenth century there was much argument about whether or not fermentation could take place in the absence of living yeast cells. However, in 1897 E. and H. Buchner succeeded in preparing from yeast a cell-free juice that fermented sugars. Such an extraction is technically quite difficult. In the following investigation we will use germinating cereal grain as a source of enzyme because it is easier to extract and because we are familiar with its effects from previous investigations.

Progress in unravelling the complexities of metabolism depends to a large extent on the ability of biochemists to extract enzymes and other complex chemicals from organisms, where they are said to be *in vivo*, and experiment with them in test-tube conditions (*in vitro*). At the same time they must obtain pure specimens of an enzyme in order to study its chemical composition. The principles of extraction are simple. When it is noted that an organism or tissue is capable of performing a specific reaction, such as starch digestion, the first step is to break up the tissue, crush or macerate it in water, and see if this mixture is capable of carrying out the reaction. If it is, then it may be filtered or centrifuged to obtain a cell-free extract. At each stage the power of the extract to perform the reaction is compared with that of the original tissue.

Unfortunately, the majority of enzymes will not stand up to extraction in this crude form and we must usually employ finer and more elaborate techniques. However, the starch-digesting enzyme of germinating barley grains is not delicate. It will remain potent in a wide range of conditions and is consequently suitable for elementary investigations, although it is not typical of enzymes in general.

Procedure

Use 8 g of dry barley grains to make 100 cm³ of extract. For each experiment you will require 10 cm³ of extract.

1   Immerse a quantity of barley grains in a solution of either sodium hypochlorite (1 per cent) for three minutes *or* mercury(II) chloride (0·1 per cent) for one minute, to surface-sterilize. Mercury(II) chloride is highly *poisonous* – be careful! Pour away the solution and wash the grain twice with sterile water.

2   Cover the grain with sterile water and leave for 24 hours at 25° C. (This temperature is recommended but not essential.)

3   Pour off the water and arrange the grain in a thin layer on wet filter paper. Cover to prevent drying out and leave for a further 24 hours at 25° C.

4   The barley grains should now appear swollen, each with a few roots protruding. Crush or grind the grains in distilled water. You can do this with mortar and pestle or, better still,

a mechanical homogenizer. With the latter see that the grains and blades are covered with water. (When using any machine, first read the instructions provided by the manufacturer.) Continue grinding or homogenizing until no intact grains remain. (See figure 61.)

5   If necessary, add water so that for every 8 g of dry barley grains used, 100 cm³ of extract is prepared.

6   Test this crude extract to see if it can digest starch. Take a small sample and add it to 1 cm³ of standard starch suspension (1 per cent). After about 10 minutes, test by adding one drop of standard iodine solution (see 5.1).

7   Filter the extract to separate dissolved enzyme from solid debris. Drape three or four thicknesses of muslin or nylon stocking across the mouth of a beaker. Slowly pour the extract into the centre. Gather up the corners of cloth and twist so that most of the fluid is squeezed out. Discard the solid debris. Test a sample of this extract with iodine solution (see stage 6).

8   Pour the milky fluid into centrifuge tubes, being careful to put equal volumes into each tube. Check that the tubes are properly balanced in the centrifuge. Spin at $1000 \times g$ for 5 minutes or for a shorter period if higher speeds are possible. Pour off the clear liquid; discard the solid material.

9   Test a sample of the extract as before (see stage 6) and store the remainder in a stoppered vessel at 4° C (domestic refrigerator temperature).

Questions

a   Which parts of the germinating barley do you think have been discarded in the extraction process?

b   What tests could be profitably performed on the discarded material?

c   Of what use would results of these tests be?

6.2 The course of an enzyme-controlled reaction

We now know that an enzyme extract is capable of digesting starch but we have little detailed knowledge of the nature of the reaction – whether it begins slowly and speeds up or whether it is influenced by temperature or other factors. We need to devise a technique which will enable us to follow the course of starch digestion so that we can obtain information of this kind experimentally.

Starch forms a coloured compound with iodine. The more starch there is, the deeper the colour produced. Thus a method of measuring the colour intensity of this compound will enable us to measure the amount of starch present. We can do this by making standard solutions of starch and iodine and comparing these visually with mixtures under test. For example, we could add one drop of iodine to a quantity of

Figure 61
A flow diagram of the extraction of amylase.

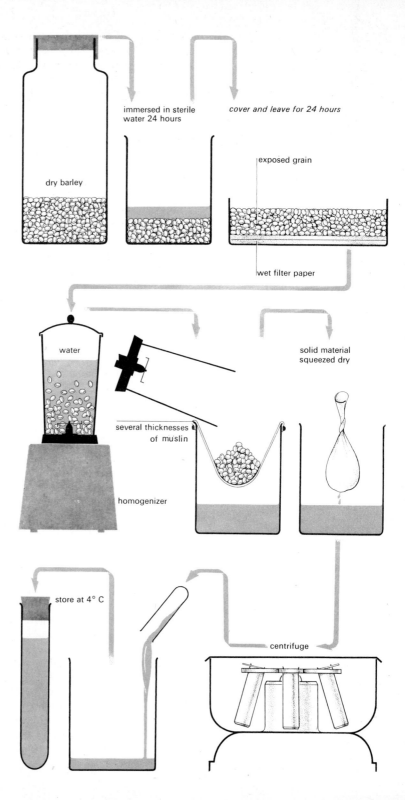

immersed in sterile water 24 hours

cover and leave for 24 hours

exposed grain

dry barley

wet filter paper

water

solid material squeezed dry

several thicknesses of muslin

homogenizer

store at 4° C

centrifuge

starch suspension and from this, prepare half, quarter, and eighth dilutions. However, visual comparison with such standards is limited by the ability of the eye and the source of illumination. As individuals we vary in our ability to distinguish between colours and our performance may change with time.

It is far better to use an instrument which measures the intensity of coloured compounds, such as some form of *colorimeter*. You may use an accurate, sensitive, and complex colorimeter or a simpler home-made version; either will provide useful information. Colorimeters of all kinds have certain features in common. Light passes through the coloured specimen under test onto a photo-sensitive element. The denser the colour, the smaller the amount of light falling on this element and the smaller the reading on a meter scale. The amount of starch present in a sample can be measured as a reading on a meter unaffected by the operator's colour vision and subjective judgment.

Figure 62
Diagrams of colorimeters.

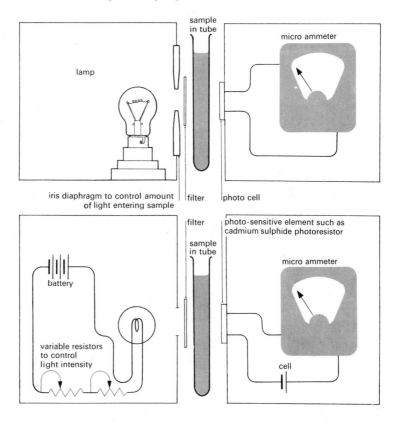

An important part of the colorimeter is the light filter and its function should be understood. Blue light passes through a blue solution, red and yellow light are absorbed by it. Thus blue light is the least useful when trying to measure the degree of blueness of the solution; red light is most useful. For this reason a red filter is used in a colorimeter when measuring the density of starch/iodine samples.

Before starting the following procedure make yourself thoroughly familiar with the workings of the colorimeter and make sure that you can get consistent readings for an unchanging sample of starch/iodine.

Procedure

Note: The instructions apply to a colorimeter capable of holding 25 cm³ of sample. If necessary, all volumes given should be modified to suit the particular colorimeter used.

1   Pour 20 cm³ water into the sample tube and add 1 cm³ standard iodine; swirl to mix well.
2   Insert the tube in the colorimeter and adjust to obtain a full scale deflection of the meter. (In some instruments such a deflection registers zero. This is known as the zero optical density of the sample.)
3   Prepare a dilute substrate by mixing 10 cm³ of standard 1 per cent starch suspension with 20 cm³ of water in a beaker. Place this in a waterbath at 15–18° C.
4   Pour exactly 10 cm³ of enzyme extract into a test-tube and put this in the same water bath as the vessel of substrate.
5   Wash out the sample tube and put in dilute iodine solution as in stage (1). Look at figure 63.
6   You are now ready to begin the experiment itself. Because it depends on time everything must be correct from the start. There will be no opportunity to rectify mistakes during the following stages. Add the 10 cm³ of enzyme extract to the 20 cm³ of dilute starch substrate, mix quickly and thoroughly and replace the beaker in the waterbath. Start a stopclock.
7   Between a half and one minute later withdraw exactly 1·0 cm³ from the reacting mixture of enzyme and substrate and add it to the iodine solution in the colorimeter sample tube. Swirl to mix. Record the time by the stopclock and the reading on the colorimeter.
8   As soon as this reading has been taken, discard the contents of the sample tube, wash it and put in a fresh iodine solution as in stage (1). At 2·5 minutes from the beginning take another sample of 1·0 cm³ from the reacting mixture and obtain a meter reading as in stage (7).
9   Continue taking samples and measuring the intensity of colour of starch/iodine at intervals such as 5, 10, 20, and 30 minutes from the start. Record your readings.
10 Perform a Benedict's test on a sample of the remaining starch/enzyme mixture.

11   Convert meter readings into units of starch concentration as follows. Repeat stage (1) and add 1·0 cm³ of dilute starch substrate (3). Obtain a reading for this mixture, for a half-strength mixture (diluted with an equal volume of water), and for one quarter and one-eighth mixtures. Plot a graph of readings (vertical axis) against starch concentration (horizontal axis).

12   Use this graph to convert the first series of readings into equivalent starch concentrations. (See figure 64.)

Questions

a   Draw a graph of starch concentration (from the conversion described above) against time and compare it with figure 65. In what ways are the two graphs similar? How do they differ?

b   Look up the logarithm (base 10) of each value of starch concentration and then plot log concentration against time. What information does the graph provide that the previous one did not reveal?

c   Name one phenomenon, not necessarily biological, which shows a similar relationship with time. Is there a similar underlying principle related to starch digestion and the phenomenon you have named?

d   How would you modify the above procedure to investigate

Figure 63
Flow diagram of the procedure for following the course of an enzyme reaction.

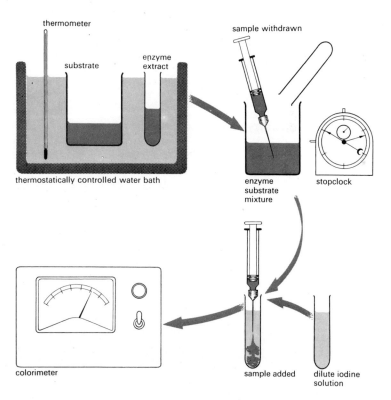

thermometer

sample withdrawn

substrate

enzyme extract

thermostatically controlled water bath

enzyme substrate mixture

stopclock

colorimeter

sample added

dilute iodine solution

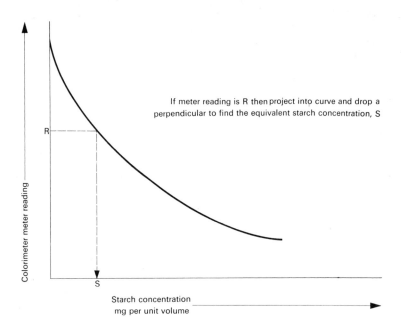

If meter reading is R then project into curve and drop a
perpendicular to find the equivalent starch concentration, S

R

Colorimeter meter reading

S

Starch concentration
mg per unit volume

Figure 64
Use of a conversion graph.

either the effect of temperature (0 to 100° C) or the effect of
acidity and alkalinity on starch digestion? Give sufficient
experimental details to serve as adequate instructions.

e   Taking into consideration the results from the tests using
iodine solution and Benedict's solution, state exactly what
reaction has taken place in the mixture of starch and enzyme
extract.

6.3 Enzyme activity in whole organisms

When you have successfully extracted an enzyme or enzymes
from germinating barley, it is appropriate to ask what function these perform in the normal life of the plant. Figure 66
shows that cereal grain is composed of the plant embryo and
a relatively large amount of starch. Starch is an insoluble
substance and unless it is digested to a soluble carbohydrate
it cannot be utilized by the embryo when it grows. Digestion
is not, therefore, a prerogative of animals. The difference
between animals and plants, in this connection, is that animals ingest starch before they digest it. Plant synthesize
their own starch, and you will understand better how this insoluble material arrives in the grain after investigation 6.4.

Starch consists of exceedingly large molecules made from
thousands of glucose units. The structural formula of a glucose molecule is shown in figure 67. The molecules may be
united in unbranched chains to form amylose (figure 68) or
as branched chains to form amylopectin (figure 69). When

Figure 65
The progress of the digestion of a protein (casein).
After Baldwin, E. (1947) Dynamic aspects of biochemistry, *Cambridge University Press* and *Baldwin, E.* (1967) The nature of biochemistry, *Cambridge University Press.*

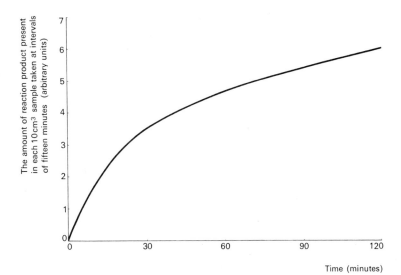

mixed with iodine solution, amylose gives an intense blue colour but amylopectin gives a dirty violet colour. This may throw some light on your observations in investigation 5.1. Natural starch molecules consist of an intimate interconnection of amylose and amylopectin, possibly something like that shown in figure 70.

When starch is digested it reacts with water (a hydrolysis reaction) forming maltose. The reaction is described in figure 71. It is likely that when a whole starch molecule is digested the amylose chains are digested before the amylopectin. This would account for the intermediate violet

Figure 66
Diagram of a longitudinal section through a cereal grain.

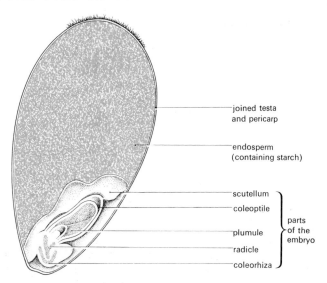

joined testa and pericarp

endosperm (containing starch)

scutellum
coleoptile
plumule
radicle
coleorhiza

parts of the embryo

colour observed in iodine tests on starch when it is being digested.

The hydrolysis of starch does not take place spontaneously or, if it does, it is so slow that it is impossible to observe. In the presence of the enzyme, amylase, the reaction proceeds swiftly; enzymes act as catalysts. Single, independent reactions like starch hydrolysis are rare in living organisms.

Figure 67
Structural formula of β glucose.

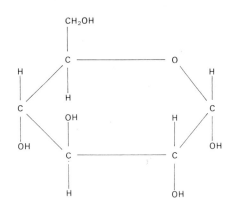

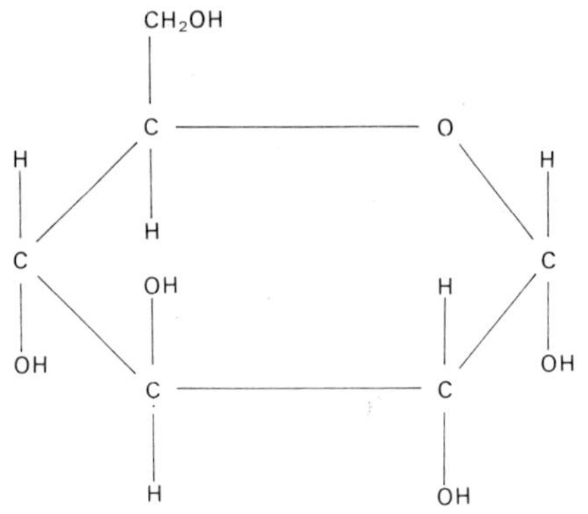

Figure 68
Structural formula of part of an amylose molecule.

(symbols C and H and most symbols O have been omitted for simplicity)

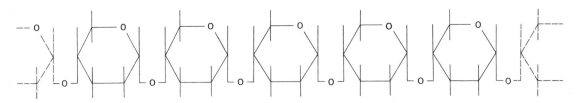

Figure 69
Structural formula of part of a branching train of amylopectin.

(symbols C and H and most symbols O have been omitted for simplicity)

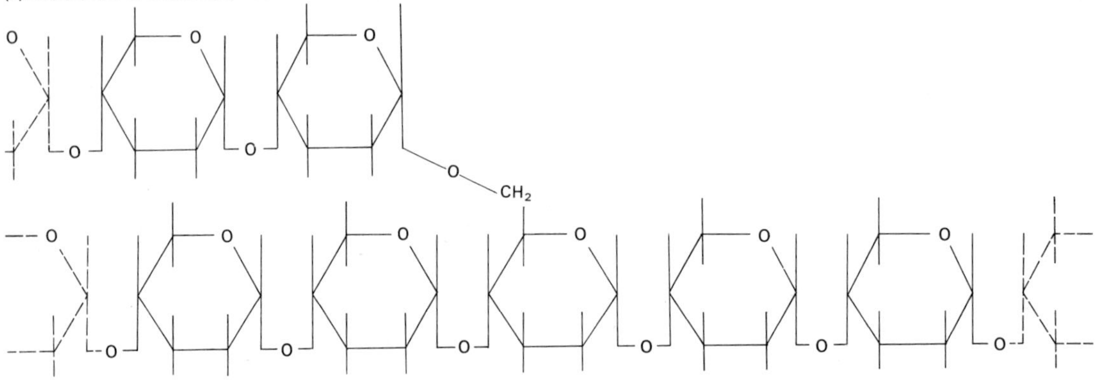

Figure 70
A diagram of a hypothetical starch
'molecule'.

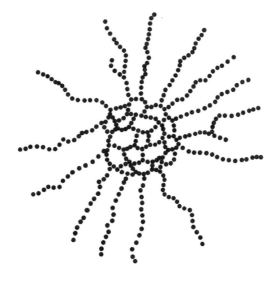

● each dot represents one glucose unit

Figure 71
Digestion or hydrolysis of amylose.

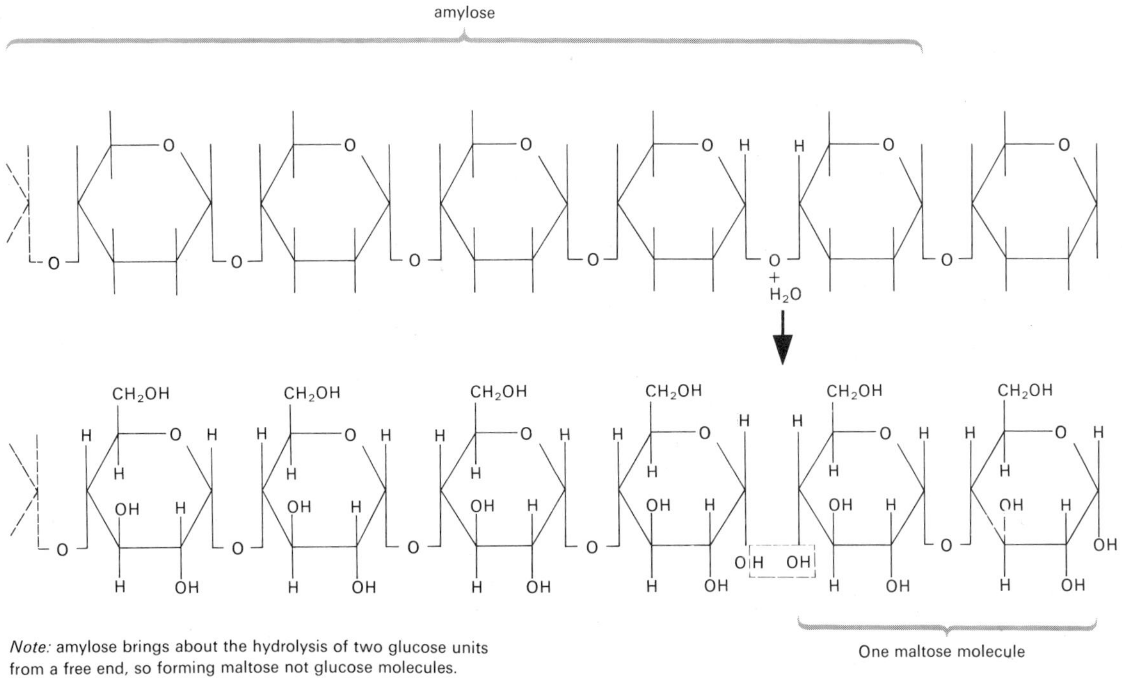

amylose

One maltose molecule

Note: amylose brings about the hydrolysis of two glucose units
from a free end, so forming maltose not glucose molecules.

It is more common to find enzymes working in systems in which the product of one reaction acts as the substrate for the next. By such closely linked series of reactions organisms break down and build up large molecules and carry out other chemical changes far more complex than the breakdown of starch to sugar.

One example of a series of linked reactions occurs when yeast ferments sugars. This is an interesting process because it plays an important part in industrial practices such as baking and brewing. In the following investigation we shall compare the system as a whole with the single reaction examined in the previous investigation (6.2). Fermentation is chosen because the products are easy to detect. The purpose of the investigation is to see if we can obtain evidence to support the hypothesis that the process is influenced or controlled by enzymes. We can do this by comparing fermentation with starch digestion. A second hypothesis to bear in mind is that fermentation consists not of one but a series of

Figure 72
A respirometer.

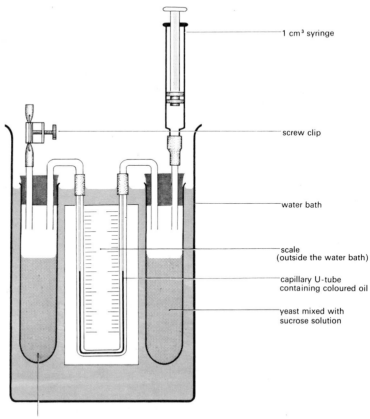

1 cm³ syringe

screw clip

water bath

scale
(outside the water bath)

capillary U-tube
containing coloured oil

yeast mixed with
sucrose solution

water in compensating vessel

reactions. It is more difficult to obtain supporting evidence for this hypothesis.

Procedure

1 Set up a respirometer. (See figure 72.) It consists of a tube containing the fermenting mixture connected to a U-tube or manometer. This, in turn, is connected to a second container which acts as a thermo-barometer, compensating for changes in temperature or barometric pressure during an experiment.

2 Take 25 to 30 cm³ of sucrose solution (containing 150 g of sucrose per litre of water) and add yeast, stirring gently, until the mixture has the consistency of cream.

3 Pour the mixture into the tube of the respirometer as shown in figure 72 so that it occupies about two-thirds of the volume of the tube.

4 Put an exactly equal volume of water in the other tube.

5 Allow a free passage of air in and out of both tubes. Fix them in a temperature-controlled water bath at 25° C. (Reliable results can be obtained only in a constant temperature even if we add a compensating vessel.)

6 Partly fill the manometer with kerosine (domestic paraffin) well coloured with Sudan III. The simplest way is to invert the U-tube so that one end dips in the oil. Suck through the other end until the required amount of oil is in the tube.

7 Connect the manometer to the tubes, as shown in figure 72, still allowing free passage of air to the tubes. Check that both menisci of the oil are at the same height.

8 When the vessels and their contents have been standing long enough to acquire the same temperature as the waterbath, push the piston of the 1 cm³ syringe to the 0·5 cm³ mark and insert it as shown in figure 72. Close the tap in the other tube so that from now on increase or decrease in gas within the vessels will affect the manometer oil.

9 Record accurately each of the following:
The position of the syringe piston (against its scale).
The height of *both* menisci (against the manometer scale).
The temperature of the waterbath.
The time.

10 Watch the movement of manometer oil and record its height in both arms of the U-tube at suitable intervals (1 to 5 minutes). Obtain at least four sets of readings before the oil reaches the top of one of the arms of the U-tube.

11 Plot a graph of the differences in height between the left and right menisci against time. Then adjust the syringe piston to bring the oil back to its original position. Note the new position of the syringe piston. This enables you to calculate the actual volume of gas given off or absorbed by the fermenting mixture.

12 If the graph is a straight line this shows that gas is being

evolved or absorbed at a constant rate. If not, the uneven rate may be due to fluctuations of temperature. Repeat stages (10) and (11) until you obtain a straight line.

13   Now that you have a means of measuring the rate of fermentation, devise an experiment to see if the process is enzyme-controlled. Change one condition of the yeast/sucrose mixture which you know affects enzyme action and note any corresponding change in the rate of fermentation.

Questions

a   Describe the experiment devised at stage (13) and state the results obtained. Do these provide evidence supporting the hypothesis that the action of yeast on sucrose is enzyme-controlled?

b   How would you determine the composition of the gas evolved?

c   How would you identify the other products of fermentation? What are they?

d   Using chemicals and apparatus mentioned previously in this chapter, how would you discover if there was a change in the amount of sucrose during the course of fermentation?

e   Design an experiment, not necessarily using a respirometer, to show whether one or a number of enzymes are concerned in fermentation. You will require a wider knowledge of enzyme properties than that derived from the work of this chapter alone.

6.4 An enzyme-controlled synthesis

Amylase and other digestive enzymes are easy to obtain and their action is simple to understand. However, you should not think that all enzyme action is as simple nor, particularly, that it is only associated with breakdown reactions. Indeed, the most important and outstanding activities of organisms, such as growth and reproduction, involve syntheses which are influenced by enzymes. No practical investigation of enzyme action would be complete without reference to at least one enzyme-controlled synthesis.

After dealing with the digestion of starch, it is appropriate to ask how starch is formed in the first place. We know it is produced in green leaves and accumulates in plant storage organs like potato tubers. Because of its relatively insoluble form (see 4.4) it is unlikely that starch is transported through the plant and it may well be that simple sugars translocated through the potato plant from the leaves are converted to starch in the tubers. Such a reaction would be the reverse of that studied in 6.1. Early attempts at getting such a reaction to take place *in vitro* failed, but in 1940 an enzyme was extracted from potatoes which produced remarkable results on a substrate closely related to simple sugars. Having practised one method of enzyme extraction and perfected a technique for measuring starch concentration, we are well equipped to attempt an investigation of this potato enzyme.

The extraction process must aim at greater purity than that in investigation 6.1 for one obvious reason; potatoes are full of starch and this is the material we wish to see synthesized. It is essential that the enzyme extract should be entirely free of starch. (In 6.2 we were concerned with starch breakdown so that its presence in the amylase extract was not a critical factor.) The following procedure investigates the synthesis of starch from glucose: 1: phosphate. (See figure 73.)

Procedure

1 Prepare a solution of 0·5 g of glucose : 1 : phosphate, dissolved in 50 cm³ of water. This will be sufficient for several experiments. The compound is unstable in solution, hydrolysing fairly rapidly to glucose and phosphoric acid at room temperature. It should be dissolved just before use or stored in a refrigerator.

Figure 73
Structural formula of glucose:1: phosphate.

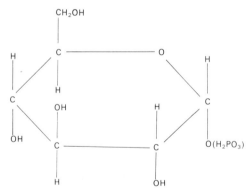

glucose:1:phosphate
(the method of numbering carbon atoms is given below)

2   Prepare other substrate solutions of the same concentration i.e. 1 per cent of glucose, maltose, or sucrose.

3   To prepare an enzyme extract for one experiment take two medium-sized potatoes, peel, cut into small pieces, and crush with a pestle and mortar or use a mechanical liquidizer. Add water sparingly so that the resulting mash is just liquid enough to be poured from its container.

4   Pour the crushed potato quickly through a single layer of muslin or stocking.

5   Transfer the extract to centrifuge tubes and spin at the highest speed ($2500 \times g$ for one minute or $1000 \times g$ for 5 minutes) to throw down the starch.

6   When the centrifuge head has stopped spinning, take one drop of clear liquid from the top of each tube and test with a drop of iodine solution on a white tile or in a test-tube. If the blue colour characteristic of starch appears, centrifuge for a further period of 1 to 5 minutes.

7   Repeat this operation until no starch is detectable in samples of the clear liquid.

8   Take three 5 cm³ samples of the clear extract and put each into a separate test-tube in a waterbath at 15–18° C.

9   Put 5 cm³ of glucose phosphate solution into a test-tube and place it in the same waterbath. Do the same with prepared solutions of glucose, maltose, or sucrose.

10   Pour 20 cm³ of water into a sample tube of a colorimeter. Add 2 cm³ of standard iodine solution and swirl to mix well. (These quantities should be modified to suit the volume of sample tube actually used.)

11   Set up the colorimeter (using a red filter) as before. (See 6.2.)

12   Add the 5 cm³ of potato enzyme extract to each vessel containing substrate. Mix each well and start a stopclock.

13   Take a 1 cm³ sample from one of the mixtures and add it to the iodine solution in the sample tube. Set the colorimeter to full scale deflection even though the mixture may be slightly turbid. This provides a zero reading for future measurements. Discard and prepare fresh iodine solution in the sample tube (2 cm³ iodine: 20 cm³ water).

14   After two or three minutes take a single drop of the glucose phosphate/enzyme mixture and mix it with a drop of iodine solution on a white tile. Look for a colour change. If there is none, repeat the operation at five and ten minutes as recorded by the stopclock.

15   As soon as the slightest trace of blue or grey colour appears on the tube, take exactly 1·0 cm³ of the glucose phosphate/enzyme mixture and add it to the iodine solution in the sample tube. Mix well and record the meter reading and time.

16   Using fresh, diluted iodine solution carry out similar measurements with the other two enzyme/substrate mixtures.

17   Check the amount of enzyme/substrate mixture left and

arrange to take 1 cm³ samples at regular intervals for the rest of the time available.

18 For each measurement made, record the name of the substrate, time, colour formed with iodine, and meter reading.

19 Convert the meter readings into starch concentrations, using the conversion graph described in 6.2 and figure 64.

20 Construct graphs for the progress of each reaction, plotting starch concentration (vertical axis) against time (horizontal axis).

Questions

a Describe the appearance of the mixtures at the end of the investigation. Which of the substrates produced starch after the addition of potato enzyme extract?

b Compare the graphs obtained in this investigation with that of starch digestion (6.2). Describe any similarities and differences between them.

c State the difference which you consider most significant and suggest a hypothesis to account for it. How would you test the hypothesis?

d It has been assumed throughout that the reaction investigated is controlled by an enzyme in potatoes. How would you test the hypothesis that an enzyme or enzymes are responsible for starch synthesis?

e Make a list of likely substrates for the potato enzyme which might be tested in addition to glucose, maltose, and sucrose.

Bibliography

Baldwin, E. (1962) *The nature of biochemistry*. Cambridge University Press. (Contains a short but stimulating account of the discovery of fermentation enzymes.)

Bell, D. J. (1948) *Introduction to carbohydrate biochemistry*. 2nd edition. University Tutorial Press. (A concise account of the commoner carbohydrates with information on fermentation.)

Eggleston, J. F. (1970) Nuffield Advanced Biological Science Topic Review *Thinking Quantitatively I, Descriptions and models, II, Statistics and experimental design*. Penguin.

McElroy, W. D. (1961) *Cellular physiology and biochemistry*. Prentice-Hall. (A slightly broader account than Baldwin's.)

Stoneman, C. F. (1970) Nuffield Advanced Biological Science Topic Review *Metabolism*. Penguin.

Synopsis

1 The use of starch production by leaves as a means of detecting photosynthesis is examined critically.

2 The influence of light on the uptake of carbon dioxide is investigated using a radioactive tracer.

3 The evolution of oxygen by the pond weed *Elodea* is a measure of the rate of photosynthesis and can be used to determine the quantitative effect of light intensity.

4 Photosynthesis occurs in the leaves of plants. Their structure is examined to see how it is adapted to this function.

Chapter 7

Photosynthesis

Summary of practical work

section	topic
7.1	The production of starch by leaves
7.2	The uptake of carbon dioxide
7.3	The evolution of oxygen
7.4	Leaf structure
7.5	Leaf pigments

7.1 The production of starch by leaves

A series of chemical changes in plants, termed photosynthesis because it depends on light, is of the utmost importance to all forms of life. Carnivorous animals depend on herbivores for food and these, of course, depend on plants. Ultimately they all depend on the one vital process of photosynthesis. For this reason a knowledge of photosynthesis is necessary for working out improved methods of food production and understanding the ecological relations of organisms.

Plants are complex organisms consisting largely of cellulose and containing varying amounts of other carbohydrates, proteins (including enzymes), oils, nucleic acids, other organic compounds, and an assortment of inorganic molecules. Quantities of these materials increase as a plant photosynthesizes and grows, and this can be shown by sampling a population and measuring the dry weight of plants from time to time. But it is far more convenient to choose one of these products and to detect photosynthesis by its presence. Traditionally, we take starch production as the criterion by which we identify photosynthesis. The reasons for choosing starch, rather than one of the other compounds mentioned above, are first, that the starch in plants, unlike cellulose and other substances, is known to change rapidly and, secondly, that starch is easy to detect. However, there are drawbacks. For example, some plants, such as grasses, never produce much starch and some plants have organs, e.g. potato tubers, that form starch in total darkness and cannot be considered as photosynthesizers. Therefore, starch cannot *always* be relied upon as an indicator of photosynthetic activity.

Consider the findings of investigation 6.4 as you carry out the following investigation.

Procedure

1 Obtain healthy specimens of one of the following types of plant, enchanter's nightshade (*Circaea lutetiana*), tobacco (*Nicotiana* sp.), geranium (*Pelargonium* sp.) or busy Lizzy (*Impatiens balsamina*). Cut a few discs with a cork borer from the leaves.

2 Test these leaves for starch as follows:
Immerse the discs in boiling water for a few seconds to break down cell walls and make them permeable.
Transfer to boiling ethanol (in a waterbath) until all the pigments have been removed.
Dip the discs in water for a few seconds to remove the ethanol. Immerse in iodine solution with potassium iodide. A blue or black colour indicates the presence of starch.
Provided the plants are healthy and have been well illuminated recently, starch should be present.

3 Put the plants in a light-proof box or cupboard for 48 hours. Again cut discs and test for starch. If none is present proceed with the next stage. If starch is present, return the plant to darkness for 24 hours and test again.

4 Cut discs from the leaves of the de-starched plant and use them to see if they are capable of producing starch in the dark as well as in the light. Float 5 to 10 discs, the right way up, on the surface of 5 per cent glucose solution in each of two small vessels open to the air. (See figure 74.)

5 Put the same number of discs into each of two similar vessels but remove the air and tap or shake the vessel until the discs sink. Fill the vessel completely with glucose solution and seal by immersing the nozzle in glucose solution and closing the tap. (See figure 74 for a method of evacuating air.)

6 Put one vessel from stage (4) and one from stage (5) in the light. Put the other two corresponding vessels in the dark. Leave for 24 hours.

7 Test the leaf discs for the presence of starch.

Questions

a   In which conditions did leaf discs produce starch?
b   What are leaf discs deprived of when they are made to sink (see stage 5)?
c   Propose a hypothesis to account for the results. How could you test it?
d   Do the results of the investigation affect the validity of the starch test as an indicator of photosynthetic activity? If so, give details.

7.2 The uptake of carbon dioxide

If we look back to the work of Chapter 1 we find that one way of detecting chemical activity in an organism is to look for changes in the composition of the immediate environment. In investigation 1·3 we analysed air samples from vessels containing locusts and *germinating* seeds. If time permits, it is well worth while carrying out this investigation with a vessel containing plants exposed to light, and to use the technique described in investigation 1·4, using small plants instead of germinating seeds. Both techniques will indicate that the presence of illuminated plants causes a rise in

Figure 74
Starch production in light and darkness: a method of setting up the experiment.

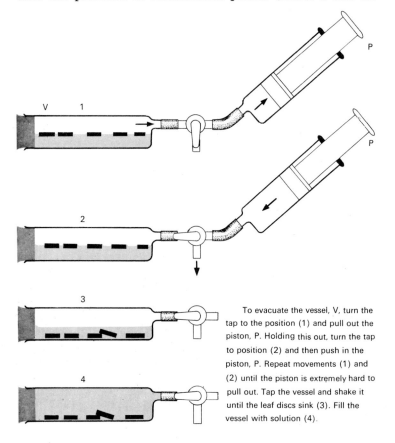

To evacuate the vessel, V, turn the tap to the position (1) and pull out the piston, P. Holding this out, turn the tap to position (2) and then push in the piston, P. Repeat movements (1) and (2) until the piston is extremely hard to pull out. Tap the vessel and shake it until the leaf discs sink (3). Fill the vessel with solution (4).

oxygen concentration and a fall in the amount of carbon dioxide in the surrounding atmosphere. Either of these phenomena provides a means of detecting photosynthesis. We consider oxygen production in investigation 7·3, that of carbon dioxide in this investigation.

When illuminated plants are enclosed, there is a detectable decrease in the carbon dioxide of the air. It is reasonable to suppose that it has been taken up by the plants, but the evidence is only circumstantial. We cannot see the gas entering the plants. In this and many similar situations in experimental biology, use has been made of 'labelled' elements. The best known of these are certain radioactive isotopes of common elements. Its radioactive property enables the isotope to be traced within an organism, yet it does not interfere with the organism's metabolism. The element behaves in exactly the same way as the non-radioactive isotopes.

By using radioactive carbon we can make labelled carbon dioxide and then see if it is taken into plants. Its radioactivity will make it distinguishable from the carbon already present in the plants. Ordinary carbon consists mostly of an isotope having an atomic weight of twelve (^{12}C). There is a radioisotope (^{14}C) which emits beta particles (electrons) that can be detected by a Geiger-Müller tube or by photographic film. This type of radiation is not a hazard to health so long as none of the radioactive material gets into the body by mouth, open wound, or other means. To prevent this happening there are strict rules for the use of radioactive tracers, which should be understood and obeyed. Unlike a poison or explosive, radioactive chemicals cannot be 'neutralized' or made safe. Some decay rapidly to a non-radioactive form but ^{14}C goes on radiating almost indefinitely.

Laboratory rules for using radioactive carbon compounds

1. Remove from the laboratory all unnecessary books and personal belongings which are not essential to the experiment.
2. Never eat or drink anything in the laboratory.
3. Always wear a laboratory coat in the laboratory (in the event of a spill, only one garment need then be sacrificed).
4. Never use your mouth or tongue when using pipettes or wash-bottles, or when fixing labels.
5. Transfer radioactive material over a plastic or metal tray.
6. Wear rubber or plastic gloves but, *even so*, never undertake work with radioactive substances when you have any kind of wound below the elbow.
7. All gloves and apparatus which may been in contact with radioactive material must be carefully checked for the presence of radioactivity (monitored) with a proper survey

instrument such as a Geiger-Müller tube and ratemeter. *All equipment must be placed in special, properly labelled containers at the end of the session.*

8. Radioactive waste (such as plant tissue which has been exposed to $^{14}CO_2$) must *not* be thrown into ordinary waste bins or sinks. It should be placed in a special container as directed by the teacher.

Procedure

1   Set up the apparatus, shown in figure 75, in a fume cupboard or similar enclosed space.

Radioactive carbon dioxide is most conveniently made by adding dilute acid to sodium carbonate.

$2HCl + Na_2{}^{14}CO_3 = 2NaCl + H_2O + {}^{14}CO_2$

The radioactive carbonate is best obtained in tablet form and the apparatus is designed for this. It can be used equally well for sodium carbonate solution.

2   Use a shoot of variegated leaves if you wish to increase the scope of the experiment to involve leaf pigments as well as light. Paper or celluloid masks may prevent the entry of gas into the leaves so transparent 'masks' serve as controls.

Under the supervision of the teacher, insert a tablet of radioactive sodium carbonate into the vessel (V) and seal it in place with a bung (B).

3   Slowly press the plunger of the syringe (S) to introduce acid. Stop as soon as the tablet has completely dissolved.

4   Switch on a source of bright light, e.g. a 150 watt bulb about 50 to 60 cm above the apparatus.

5   Close the fume cupboard, switch on the ventilating system, put up a sign 'danger radioactivity', and leave the apparatus for 48 to 72 hours.

6   Switch off the light, open the fume cupboard, but keep the ventilating system running. Remove the syringe containing acid so that the tube (T) is open to the air.

Join a filter pump to the tube (P) so that air is swept through the apparatus into the flask of potassium hydroxide solution. Leave this running for at least three hours.

7   Monitor the equipment with a survey meter and remove the bell jar.

8   To prepare for autoradiography of the leaves first cut a piece of cardboard to the same size and shape as the X-ray film to be used, approximately 12×16 cm. Draw lines dividing the card into four parts each large enough to accommodate a leaf, and write in each a description such as 'leaf with black mask', 'leaf with transparent mask'.

9   The leaves are now to be removed from the shoot and pressed against the X-ray film for at least three days. Removal will not immediately kill them and any carbon taken in may be respired away unless all chemical action is stopped at the moment the leaves are plucked. One method is to use a fixative as follows (see figure 76).

Figure 75
Apparatus for exposing plants to
radioactive carbon dioxide.

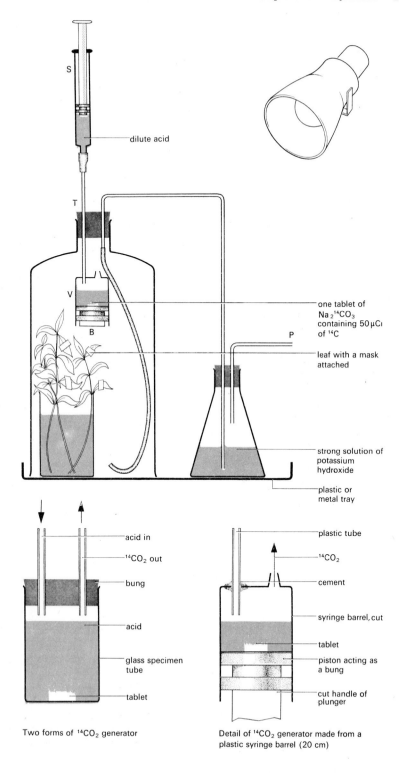

S

dilute acid

T

V

B

one tablet of
Na$_2$^{14}CO$_3$
containing 50 μCι
of ^{14}C

P

leaf with a mask
attached

strong solution of
potassium
hydroxide

plastic or
metal tray

acid in

^{14}CO$_2$ out

bung

acid

glass specimen
tube

tablet

plastic tube

^{14}CO$_2$

cement

syringe barrel, cut

tablet

piston acting as
a bung

cut handle of
plunger

Two forms of ^{14}CO$_2$ generator

Detail of ^{14}CO$_2$ generator made from a
plastic syringe barrel (20 cm)

Put a large smear of polystyrene cement in the centre of each rectangle on the card. Using forceps, remove the masks. Then pick off the leaves from the plant and lay them on the cement in their appropriate positions. With a glass rod, press the leaves into the cement gently until it is evenly absorbed. The leaves will turn a darker shade of green.

10   Allow the cement to dry and then put the whole card into a thin polythene bag and seal it with sticky tape. Mark it 'radioactive'.

11   Take the sealed card to a darkroom and, under a photographic safe light, place the card on a sheet of X-ray film in a folder (see figure 76). Close the folder and take it to the fume cupboard or some other place specially reserved for radioactive material. Place a heavy weight on top of the folder to

Figure 76
Setting up an autoradiograph.

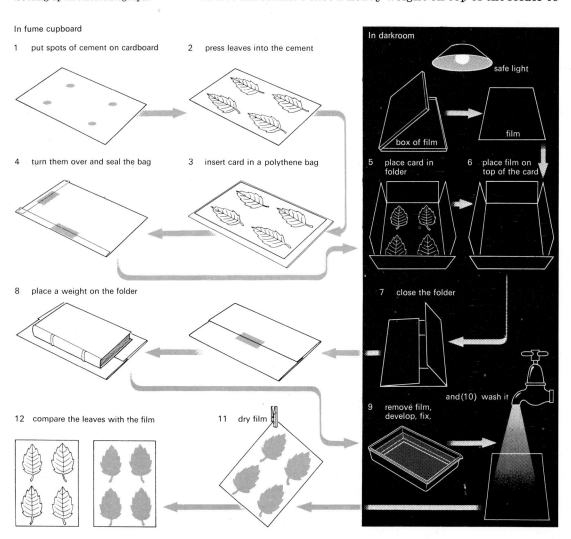

press film and card together and leave it at least three days.

12 Take the folder to the darkroom and process the film according to the manufacturer's instructions. When you have answered the questions below, dispose of the card with leaves (still in the polythene bag) as directed by the teacher.

13 Dispose of the remainder of the plant and the potassium hydroxide solution as directed. Monitor all apparatus which has been in contact with radioactive material.

14 The radioactive emission of radiocarbon (^{14}C) sensitizes photographic film so that when it is developed, silver halide is turned into grains of silver and portions of the film appear black. Examine the processed films carefully. Compare them with the cards bearing leaves. Compare the images corresponding to the masked and unmasked leaves and those which had transparent masks on them.

Questions

a What conclusions can you draw from your observations?

b Do all the blackened areas associated with leaf images correspond with your expectations?

c Examine the centre or mid-rib portion of the leaf images on the film and report your findings.

d Are there any blackened portions of the film outside the leaf positions? If so, how do you account for them?

e In what compound or compounds do you think the radioactive carbon was combined at the time of cutting leaves from the shoot?

f How would you find the answer to question (e) experimentally?

7.3 The evolution of oxygen

We know from the work of section 7.2 that the fixation of carbon dioxide in a leaf depends on light. However, from that evidence alone we do not know whether photosynthesis proceeds at a steady rate once a certain threshold of light intensity has been reached, or whether the rate is proportional to the intensity of light. This is an important problem. Its solution will not only further our understanding of the metabolic processes of plants but will be of practical value because it will provide information about the best amount of light needed to obtain maximum production of materials in plants grown for food.

We can use the evolution of gaseous oxygen as a method of investigating photosynthetic activity. The aquatic plant *Elodea* is a suitable experimental organism. You may have noticed during the work of Chapter 1, in which aquarium tanks were set up as model communities, that bubbles of gas rose to the surface from aquatic plants such as *Elodea* on sunny days. If the gas is oxygen, a product of photosynthesis, measurement of the amount produced can provide a means of assessing the rate of photosynthesis.

The capillary tube method of analysing air (see investigation 1.3) can be easily modified for the analysis of gas from *Elodea*; one such modification is illustrated in figure 77. There is no need to follow this design exactly. Any method of causing gas to collect at one site in the tube will serve adequately. The wide connecting tube between the capillary tube and the syringe provides a convenient means of disposing of gas samples once their volume (indicated by length) has been measured.

For photosynthesis to occur there must be a source of carbon. For aquatic plants this is in the form of bicarbonate ions (HCO_3^-) present in the water. Thus, when carrying out the investigation you should realize that this environmental factor may vary as well as light. An additional or alternative investigation may be carried out in which the light intensity is fixed but the concentration of bicarbonate ions is varied. If temperature is varied, then allowance must be made for changes in the solubility of oxygen in water.

Procedure	1 Prepare a glass capillary tube for collecting gas bubbles from a shoot of *Elodea*.

1   Prepare a glass capillary tube for collecting gas bubbles from a shoot of *Elodea*.

2   Put a bright light source near an aquarium tank containing *Elodea* for several hours. If no bubbles appear, add potassium bicarbonate solution sparingly to the water.

3   Choose a piece of *Elodea* which has a steady stream of bubbles coming from it. Transfer it to a test-tube or specimen tube of convenient size filled with water from the tank. Set up the apparatus shown in figure 77 (or a similar device for collecting gas).

4   Set the specimen tube in a larger vessel containing water and a thermometer so that you can check that the temperature remains constant through the following stages.

5   Put a bright light source such as a lantern slide projector close to the apparatus so that light falls directly on the piece of pond weed. Darken the room to exclude other sources of light.

6   The easiest way to change the intensity of light falling on the plant is by moving the source away. Measure the initial distance between the *Elodea* shoot and light source.

7   The volume of gas collected can be estimated (see figure 77) by drawing the bubble along the capillary until it is alongside the scale and measuring its length. Once measured, the bubble can then be drawn further along to the wide tube where it will remain conveniently out of the way during subsequent measurements. Measure the distance from the projector to the pond weed and the quantity of gas evolved in a

known period e.g. 5 to 10 minutes.

8   Double the distance between light source and plant specimen. Check that the temperature remains constant. Collect
and measure a second sample of gas in the same period as
before.

9   Continue to make such measurements with the light source
at greater and greater known distances from the pond weed.

10   The intensity of light falling on a given area, such as a piece
of weed, from a constant source, such as a projector, is in-
versely proportional to the square of the distance between
them. If, for example, the distance is doubled, the light
intensity is decreased not by $\frac{1}{2}$ but by $\frac{1}{2^2}$ or $\frac{1}{4}$. Plot a graph of
the amounts of gas collected in each equal period of time
(vertical axis) against $\frac{1}{d^2}$ (horizontal axis) where d is the
distance between the plant and the light source.

11   Place the light as near the plant as possible and collect a
large sample of gas, so that the bubble is several centimetres
long. Analyse it using solutions of potassium hydroxide and
potassium pyrogallate as described in investigation 1.3.
Record the percentage of carbon dioxide and oxygen in the
sample.

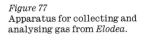

Figure 77
Apparatus for collecting and
analysing gas from *Elodea*.

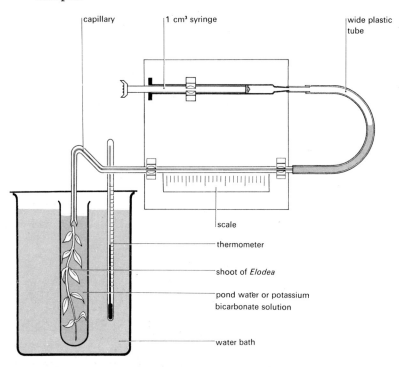

Questions

a   State any relationship between gas production and light intensity which has been demonstrated by your results.

b   How would you confirm, experimentally, that the light intensity decreases with the square of the distance?

c   What is the composition of the gas analysed? How do you account for it?

d   What changes can occur in a bubble of gas as it rises from a piece of pond weed to the surface of the water? How might these changes affect your results?

7.4 Leaf structure

The most conspicuous structural feature of plants is their leaves. Apart from a few exceptions such as those of cacti, leaves are numerous, thin, flat, green organs. It is reasonable to suppose that, as leaves are invariably exposed to light, they are the photosynthetic organs of a plant.

Though the sizes and outline shapes of leaves vary greatly from one species to another, their thickness is remarkably similar. This may be related to their photosynthetic function. Investigation 2.1 indicated that the characteristic thin leaf-form can be related, in part at least, to the speed of gaseous diffusion. If leaves were much thicker, carbon dioxide and oxygen would not be able to travel to and from the interior fast enough to sustain life. Any study of photosynthesis, however brief, should pay due attention to the organs in which it occurs. In the following investigation an attempt is made to relate the structure of leaves to their photosynthetic function.

Some leaves, like those of mosses and the pond weed *Elodea*, are thin enough for internal details to be seen under the microscope without any special treatment. Most are too thick unless previously 'cleared' (figure 78). Clearing enables cell detail to be seen at all levels but kills the leaves in the process. Thin sections of fresh leaf material are difficult, but not impossible, to cut. Better sections are obtained by embedding a leaf in wax and using a microtome to cut it. This, like clearing, kills the leaves and alters their colour and some important structural features. The best approach is to use all the techniques available, bearing in mind the deficiencies of each, to build a mental model of leaves in their natural condition.

Procedure

1   Pick a leaf from a shoot of *Elodea* and mount it in a drop of water under a cover-glass. It may be curled and therefore easier to mount if you cut it into two or three pieces with fine scissors first. Observe under medium and high power magnification. Insert an eyepiece graticule (linear).

Answer questions (a) and (b) below.

2   Examine cross sections of several leaves of land plants cut by a microtome. Note that apart from minor variations between different species there are two basic patterns of leaf organization within the whole group of flowering plants. The basic organization of long, parallel-veined leaves differs from that of broad, net-veined leaves. Look at sections of both types and attempt question (c).

3   Pick a leaf from a plant and cut transverse sections by hand. Two methods are indicated in figure 79. Do not attempt to cut complete sections.

Hand-cut sections are seldom as thin or as complete as those produced by a microtome. However the purpose of hand-cutting is not to emulate the machine but to make it possible to examine fresh material. Leaf material prepared for microtome cutting does not retain its colour.

Keep the cut leaf wet. Put the sections in a watch-glass full of water, select the thinnest, and mount them in water under a cover-glass. Observe under the medium and high power of a microscope.

Answer questions (d) and (e).

4   Mount a few thin sections in a drop of iodine solution and observe as before.

5   Cell contents which become detached from leaf sections are more clearly visible than those *in situ*. Look at the debris in the water surrounding the hand-cut sections under H.P. magnification.

Figure 78
Stomata in the abaxial epidermis of a plane tree leaf (*Platanus acerifolius*). ($\times$ 80.) The specimen has been cleared with hypochlorite bleach to remove cell contents and then stained.
Crown copyright.

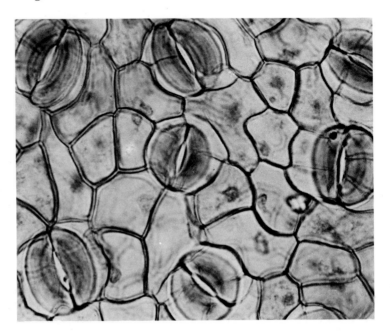

Questions

a   How many different kinds of cell, judged by their shape, can you see in a leaf of *Elodea*? Describe the size and contents of the cells.

b   How many different kinds of cell (judged by size and shape) can you see in a section of a broad leaf of a dicotyledonous plant?

Give a brief, illustrated account of each type.

c   By two simple plan drawings compare the organization of cells in a broad leaf (dicotyledon) with that in a narrow leaf (monocotyledon).

d   Describe the contents of the cells of a freshly cut leaf. Compare their colour, size, and shape with those of the contents of *Elodea* leaf cells.

e   Describe any items of debris which occur commonly and

Figure 79
Methods of cutting leaf sections.

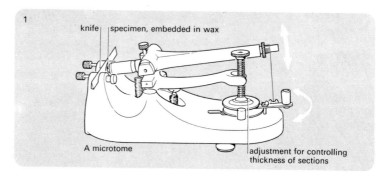

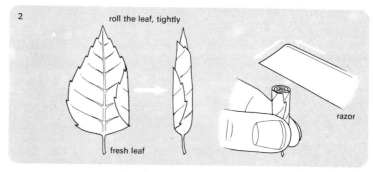

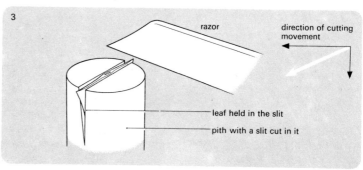

which you think are important in the whole leaf. Do those in
iodine solution differ from those in water?

f   What structural features of the leaves examined so far can
be regarded as adaptations to a photosynthetic function?

7.5 Leaf pigments

One feature of plants which is more characteristic than
leaves is colour. Most plants are green or, to put it another
way, most of the organisms capable of photosynthesis contain green pigments called chlorophylls. Presumably
chlorophyll plays a vital part in the process. Before accept-
ing such a proposition it is worth recalling that quite a large
number of plants, like the copper beech and some seaweeds,
are not green but red or brown.

To find out exactly what pigments do in the mechanism of
photosynthesis is beyond our scope here but, as a first step,
it is necessary to find out if chlorophyll is present in the red
and brown plants and if such non-green pigments are present
in other plants which appear green. When these problems
have been solved we shall then be in a better position to
make general statements about the possible connections
between pigments and photosynthesis.

The separation of substances by chromatography, already
mentioned in connection with amino acids (see 5.3), was first
demonstrated by Tswett in 1907 using extracts containing
plant pigments. The technique is simpler for pigments than
for amino acids and separation can be effected more rapidly.
Prepared thin layers can be used, as before, but paper will do
as well and has the advantage of being cheaper and easier to
handle.

Procedure

1   Collect a handful of leaves from several different plants,
some green, and some brown or variegated. Keep the collections separate.
2   Tear the leaves into small pieces and shred them in an
electric grinder.
3   Scrape the fine shreds into a large test-tube and add just
enough acetone to cover them. Shake well and allow the
mixture to stand for at least an hour. Then decant off the
clear, coloured liquid.
4   Cut strips of chromatographic paper about 5 to 6 cm by 20 to
30 cm. Handle the paper by the edges because finger marks
spoil the process of pigment separation.
5   Rule a light pencil line 2 cm from the edge along one end and
put a small drop of extract at its centre. When this has dried
put another drop on the same place and repeat the process so
that a small but concentrated spot accumulates. It should

not be wider than 3 to 4 mm. Use a wire loop or short length of thin capillary tubing to dispense the extract.

6   Prepare the solvent by adding one part of acetone to nine parts of petroleum ether (boiling range 80–100° C) and pour it into a jar to a depth of 1 cm.

7   Suspend the loaded paper vertically in the jar with the spot at the bottom but *above* the surface of the solvent. (See figure 80.) Close the container so that the paper is surrounded by air saturated with solvent vapour.

Figure 80
Setting up a paper chromatograph.

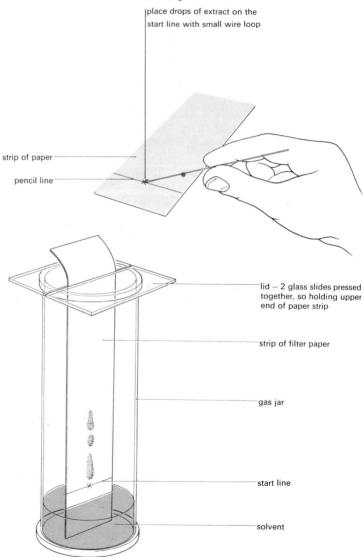

place drops of extract on the start line with small wire loop

strip of paper

pencil line

lid – 2 glass slides pressed together, so holding upper end of paper strip

strip of filter paper

gas jar

start line

solvent

8   The solvent will ascend the paper rapidly, carrying the pigment with it. After 20 to 30 minutes remove the paper and allow it to dry.

9   Repeat the procedure using the other extracts.

10   Count the number of spots on each paper. Describe the colours. As these fade quickly, outline each spot with pencil for future reference.

11   Though the distances between spots depends on duration and experimental conditions, the order from top to bottom is always the same.

Refer to a textbook or other sources of information and try to name the pigments by colour and relative position.

12   The reference may use the term Rf. This is a quotient obtained by dividing the distance through which a substance has moved by the distance through which the solvent has moved (in the same time and units). For example if one component of a mixture rose 15 cm from the start line and the solvent moved 20 cm from the same line, then for this component

$$Rf = \frac{15}{20} = 0 \cdot 75$$

Calculate the Rf values for each spot from one extract and see if these correspond with values given in a text or reference. (Bear in mind that Rf values are affected by the methods of extraction and the solvents used.)

Questions

a   Do all the pigment extracts contain the same variety of components? If not, give details.

b   Are some component pigments common to all the extracts?

c   What conclusions can you draw from answers (a) and (b) concerning the role of pigments in photosynthesis? Try to find, from textbooks, the function performed by plant pigments.

Bibliography

Baron, W. M. M. (1967) *Organization in plants*. 2nd edition. Edward Arnold. (Plant pigments and other aspects of photosynthesis.)

Bassham, J. A. (1962) 'The path of carbon in photosynthesis.' *Scientific American*, **206**, 6, 89. (An illustrated account of modern research methods.)

Gaffron, H. (1965) *Photosynthesis*. B.S.C.S. Pamphlet 24. D. C. Heath & Co. (A short but comprehensive account of photosynthesis.)

Galston, A. W. (1961) *The life of the green plant*. Prentice-Hall. (Places photosynthesis in the wider context of plant nutrition and provides an introduction to modern theory.)

Stoneman, C. F. (1970) Nuffield Advanced Biological Science Topic Review *Photosynthesis*. Penguin.

Synopsis

1 Overall metabolic activity in organisms and tissues which respire aerobically is usually related to oxygen uptake. An accurate method of measuring the oxygen absorption of small organisms is described.

2 The respiratory quotient provides a means of detecting the substances respired, but the usefulness of this ratio requires critical examination.

3 Advances in biochemistry are usually made by breaking down living tissues and isolating particular chemical systems. Two examples illustrating this technique are the chemistry of respiration and that of photosynthesis.

Extension work III
Metabolic systems

Summary of practical work

section	topic
E3.1	The uptake of oxygen as a measure of metabolism
E3.2	Respiratory quotient
E3.3	Investigation of an intermediate respiratory reaction
E3.4	Chloroplast activity in photosynthesis

The aim of the investigations which follow is to find out a little more about the way in which organisms maintain themselves in the living condition. One way, elaborated in previous sections, is to look at the internal structure of dead animals and plants, and try to deduce the function of internal parts when the creatures were alive. The chief obstacle to using living organisms in such an enquiry is that as soon as we try to examine a living process the very act of investigation so upsets the organism that it dies. We are in a situation similar to that of a man trying to find out how a motor car engine works. As soon as he removes the casing to look inside, the engine stops.

Enquiry into the internal workings of living creatures must therefore be circumspect and, in a sense, the chemical equivalent of dissection. In the first and second of the following investigations the organisms remain alive and unchanged, but in the third and fourth we examine only a small part of the living chemistry; the remainder is discarded.

E3.1 The uptake of oxygen as a measure of metabolism

In Chapter 1 of the *Laboratory Guide*, we drew attention to the uptake of oxygen and output of carbon dioxide by living organisms. Because these gases are associated with respiratory metabolism, we can learn something of the rate and nature of respiration by further study of gas exchange. To do this we require a more exact technique than the one used in Chapter 1. A modified version of that used in Chapter 6, to measure carbon dioxide evolution by yeast, will serve to measure the uptake of oxygen and overall gas changes brought about by small organisms. The apparatus (see figure 81) consists of two vessels, of which one contains the organisms and the other acts as a thermobarometer. Small changes in temperature or pressure cause air in this vessel to expand or contract, opposing and compensating for similar changes in the other vessel. Changes in the manometer are thus due only to the activities of the organisms themselves.

Figure 81
A respirometer measuring the uptake of oxygen by seeds.

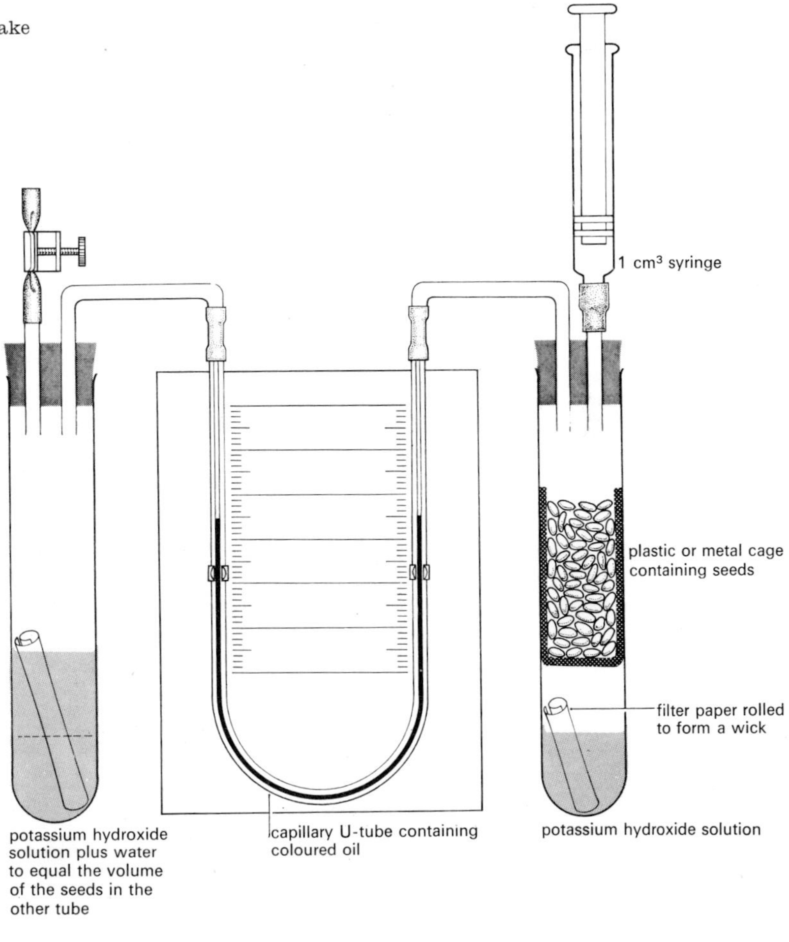

1 cm³ syringe

plastic or metal cage containing seeds

filter paper rolled to form a wick

potassium hydroxide solution

capillary U-tube containing coloured oil

potassium hydroxide solution plus water to equal the volume of the seeds in the other tube

Yeast produces much carbon dioxide and takes up little if any oxygen. If, instead, we use organisms such as germinating seeds or locust hoppers, these may absorb as much oxygen as the carbon dioxide they produce. The manometer will, therefore, record the net difference between gas uptake and output. The actual amount of one gas is found by removing the other, and the most convenient way of doing this is to remove the carbon dioxide as it is produced. Strong potassium hydroxide solution absorbs carbon dioxide readily.

Procedure

1   Pour 5 cm³ of potassium hydroxide solution (15 per cent) into both respirometer tubes; use a funnel so that none touches the sides of the vessels.
2   Add small rolls of filter paper to act as wicks.
3   Fill the basket or cage with living organisms such as germinating seeds and put it into a vessel, making sure that the seeds are not touching the potassium hydroxide solution or wick.
4   Fix the bung with a 1 cm³ syringe and connecting tube, as shown in figure 81.
5   Estimate the total volume of the seeds and basket together and add this amount of water to the other respirometer vessel.
6   Draw coloured kerosine or Brodie's fluid into the manometer tube so that it is free from bubbles and comes to about the half-way mark on the scale, on each side.
7   Open the screw clip and remove the syringe; then connect the manometer U-tube.

Figure 82
The respirometer in use.

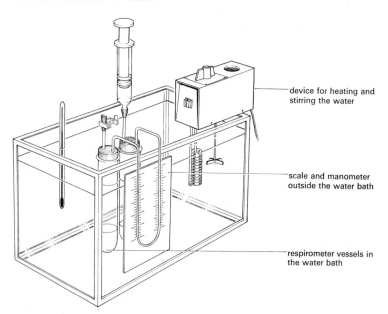

device for heating and
stirring the water

scale and manometer
outside the water bath

respirometer vessels in
the water bath

8   Place the respirometer so that both tubes are immersed in a waterbath maintained at 20° C but the manometer is suspended outside (see figure 82).

9   Set the piston of the syringe at about the 0·5 cm³ mark and, when the respirometer has been in the waterbath for 5 minutes, insert the syringe as shown. Close the screw clip. By means of the syringe adjust the manometer so that the fluid levels are equal on both sides.

10   Record the exact positions of the syringe piston, *both* manometer menisci and the time.

11   Record new positions of the manometer fluid at 4-minute intervals. When it nears the end of the scale on one side restore it to its original position and note the new position of the syringe piston.

12   Plot a graph of meniscus level against time. Continue to take readings until four consecutive ones lie on the same straight line.

13   Raise the temperature of the waterbath to 30° C and repeat stages (9)–(12).
       Remove and weigh the living material.

Questions

a   When the graph of manometer readings against time is a straight line, what can be said about the uptake of oxygen by organisms in the respirometer?

b   How much oxygen was absorbed by the organisms? State the amount as cubic millimetres (mm³) per hour per milligramme (mg) of living material.

c   How does a rise in temperature affect the rate of oxygen uptake? How can the effect be best expressed?

d   What rates of oxygen uptake would you expect if you raised the temperature to 40° or 50° C? Give your reasons.

E3.2 Respiratory quotient

By measuring the uptake of oxygen and output of carbon dioxide, we can tell if organisms are respiring and how fast. The respiratory rate, measured as the volume of oxygen taken up per unit weight of tissue per unit time, is a useful guide to the general metabolic activity of an organism. But there are difficulties associated with this kind of measurement. For example, organisms differ in the amount of non-respiring material in their bodies. Comparison between the respiratory rates of insects, with their non-respiring exoskeletons, and earthworms, with their thin cuticles, would be invalid.

It is possible, using the same apparatus as before, to discover something about the nature of the material respired. This is done by comparing the amount of carbon dioxide given off with the amount of oxygen absorbed by the same tissue in the same time. Such a comparison is simply expressed as:

Respiratory quotient (R.Q.) =

$$\frac{\text{volume of carbon dioxide given out}}{\text{volume of oxygen absorbed.}}$$

These volumes must be measured in the same units for the same period of time, keeping temperature and other conditions constant. The respiratory quotient is a ratio, independent of the amount of respiring tissue and units used. Its usefulness depends on the fact that equal volumes of gases (at the same temperature and pressure) contain equal numbers of molecules – Avogadro's hypothesis. If a hexose sugar such as glucose is completely respired, the numbers of carbon dioxide and oxygen molecules involved are the same.

$$C_6H_{12}O_6 + 6\,O_2 = 6\,H_2O + 6\,CO_2.$$

The calculated R.Q. for hexoses is thus 1·0. Fats have far less oxygen than carbohydrates per molecule, that of the fat tristearin having the atomic composition $C_{57}H_{80}O_6$. Complete respiration of this fat can be expressed as:

$$2\,C_{57}H_{80}O_6 + 163\,O_2 = 110\,H_2O + 114\,CO_2.$$

The calculated R.Q. is $\dfrac{114}{163} = 0\cdot69$.

Other examples of respiratory quotients are,

Amino acid glycine	1·33
Amino acid leucine	0·80
Amino acid lysine	0·86
Citric acid	1·33
Oxalic acid	4·00.

A glance at these figures is enough to show that it is not possible to deduce from knowledge of respiratory quotient alone that any one subtrate is being respired. R.Q. values, when used in conjunction with other evidence, are aids in the complex task of unravelling the web of metabolic reactions which constitute tissue respiration. Some difficulties and disadvantages associated with the concept of respiratory quotients are suggested by the questions which follow.

The amount of carbon dioxide produced by organisms can be determined indirectly, using the same apparatus as before but omitting the potassium hydroxide solution. If the volume of oxygen absorbed by respiring organisms is exactly equal to the volume of carbon dioxide produced, then there will be no change in the manometer. But if, for example, there is an apparent production of 0·25 cm³ of gas, using the respirometer for 20 minutes, but a decrease in volume of

0·50 cm³ in the same period when the potassium hydroxide solution is introduced, then we can deduce:

Uptake of oxygen = 0·50 cm³
Carbon dioxide produced – oxygen absorbed = 0·25 cm³
∴ carbon dioxide produced = 0·25 + 0·50
= 0·75 cm³

$$\text{R.Q.} = \frac{0·75}{0·50}$$

$$= \underline{\underline{1·50}}.$$

Procedure

1 Using a respirometer find the amount of oxygen absorbed by germinating seeds in a period of 45 minutes at 20° C. Follow the same procedure as in the previous section; manipulate the syringe to maintain the manometer fluid at a constant level.

2 Remove the potassium hydroxide solution from both vessels and wash them out with water.

3 Replace the basket containing the germinating seeds and the bungs; set the respirometer in the waterbath at 20° C. During a period of 45 minutes record any increases or decreases in gas volume.

4 From these records calculate the amount of carbon dioxide which has been produced in the period.

Questions

a Calculate the respiratory quotient of the seeds.

b What is the respiratory substrate most likely to be present in seeds? What tests could you perform to discover the nature of the substrate? Does the R.Q. correspond with your expectations?

c What would you expect the R.Q. of living yeast to be (see Chapter 6)?

d Are question (c) and its answer in any way relevant to the respiration of germinating seeds? Give your reasons.

e Suggest three different hypotheses to account for a respiratory quotient of unity (1·00) found by experiment.

f In 1905, C. R. Barnes wrote: '... This respiratory ratio [R.Q.] has proved a veritable will o' the wisp leading investigators into a bog where their labours and their thinking were alike futile ...'. Explain one use of R.Q. which you think would be futile and one which might be fruitful.

E3.3 Investigation of an intermediate respiratory reaction

When studying the structure of an organism we can gain some information by looking at its external features but far more by dissecting it and examining the internal structure. Similarly, when investigating the chemistry of living creatures, we can only derive a limited amount of information from intact organisms and usually need to extract and separate certain components and study them in isolation. When the chemical nature of several of these isolated systems is understood, their inter-relationship in whole organisms can be established. Metabolism is thus explored by a process of breaking down and building up. The following procedure shows how one stage in the chemistry of respiration can be investigated in isolation.

The word 'respiration' is used to denote a whole series of chemical reactions, taking place in living tissue, by which energy is transferred from molecules such as starch and glucose to energy-requiring processes such as synthesis, absorption, and movement. The series is a complex one, far more so than the simple oxidation implied by the equation:

$$C_6 H_{12} O_6 + 6 O_2 = 6 H_2O + 6 CO_2.$$

One of the intermediate reactions involved is the oxidation (by removal of hydrogen) of succinic acid to fumaric acid:

$$\begin{array}{ccc} CH_2{-}COOH & & CH{-}COOH \\ | & \longrightarrow & \| \\ CH_2{-}COOH & & CH{-}COOH. \end{array}$$

(Note that when such an oxidation occurs some other element or compound must be reduced; this is not shown above.) To see how these substances fit into the overall sequence of respiratory reactions you should consult a textbook of biochemistry. The practical question that immediately confronts us is, 'How can the activity of a substance such as succinic acid be studied distinct from the hundreds of other compounds present in living organisms?' Examination of the formulae, above, shows that two hydrogen atoms are removed from each molecule of succinic acid to form one of fumaric acid. These are substances which accept such hydrogen atoms and in doing so, change colour. They could obviously be useful in this context, and one example is 2:6 dichlorophenolindophenol. It is blue in its oxidized form but loses its colour when reduced (see figure 83).

Figure 83
Structural formulae of 2:6 dichloro-phenolindophenol (dicpip), showing acceptance of 2 hydrogen atoms.

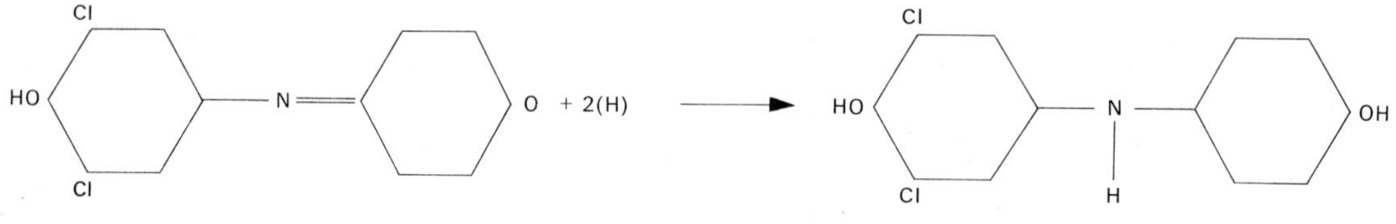

coloured form (blue in alkali, red in acid)

colourless, reduced form

If the coloured form of dichlorophenolindophenol (dicpip) is decolorized by a tissue extract, one explanation could be that it has accepted hydrogen atoms from succinic acid. This hypothesis can be tested. As most living processes are governed by enzymes, it is of interest to see if this reaction is affected by temperature, or other factors, in a manner characteristic of enzymes.

Respiratory reactions are associated with the presence of minute cell inclusions called mitochondria. If these are mixed with pure water they swell and burst, and their chemical activity decreases or stops altogether. Mitochondria can be kept intact by adding sucrose to the extract medium so as to raise its osmotic strength. They are also sensitive to changes in pH, and extraction should be performed in a buffer solution to maintain a constant pH, similar to that of intact cells. When tissues are broken up for subsequent investigation, substrates may become exhausted by the chaotic reactions that ensue. To preserve the *status quo* during extraction the temperature, and therefore the activity, of the material should be kept low. Using this information it is possible to examine the oxidation of succinic acid by colorimetric estimation of dicpip solution.

Procedure

1   Prepare a phosphate buffer solution and add sucrose. Divide it into two equal parts and to one add succinic acid. Store the solutions in a refrigerator at 0° C.

2   Take about twelve mung bean seedlings which have germinated and grown in the dark. Remove their testas and push them into cold centrifuge tubes (capacity 15 cm³) with a cold glass rod. Add 1–2 cm³ of buffer/sucrose lacking succinic acid and crush by exerting a firm pressure with the rod.

3   Fill the centrifuge tube with buffer/sucrose, mix the contents well, and then pour half into another tube. Add more buffer/sucrose until both are filled to within 1 cm of the brim and contain equal quantities of liquid. Place the tubes in opposite positions in the centrifuge head.

4   Spin the tubes at 2500 g for three minutes.

5   Pour off the clear liquid from both tubes into the sample vessel of a colorimeter.

6   Add 0·1 cm³ dicpip solution (1 per cent in sucrose/buffer) to the sample vessel and quickly mix well. Using a red filter take a reading and add more dicpip solution, if necessary, to obtain a value at about the mid-point on the scale.

7   Record this reading and the time. Take further colorimeter readings at 1, 2, 5, 10, 15, 20, and 30 minutes after the addition of dicpip solution.

8   Carry out the same procedure, using seedlings from the same batch, and make an extract using buffer/sucrose with succinic acid added.

9   *Either* plot both sets of readings as a graph; *or* convert your colorimeter readings into dicpip concentrations by making five solutions as follows:

0·2 cm³ of dicpip solution (1 per cent) in 100 cm³ of water;

0.4 cm³, 0.6 cm³, 0.8 cm³ and 1.0 cm³ of dicpip in the same quantity of water.

Obtain colorimeter readings for each and plot a graph of these against concentration. Use this as a conversion graph, and then plot converted experimental results against time.

Questions

a   Do your results support the idea that succinic acid is oxidized in an extract of mung bean seedlings? (Give your reasons.)

b   Do the results obtained when using succinic acid added in the medium further support this idea?

c   How would you find out experimentally if the oxidation of succinic acid is enzyme-controlled?

d   If dicpip is an indicator of succinic acid oxidation and this oxidation is a stage in the chemistry of respiration, do you think that dicpip solution could be used as an indicator of respiration as a whole? If so, design a simple experiment to test this idea.

e   The reduction of dicpip is not specifically connected with the oxidation of succinic acid. What steps could be taken to avoid decolorization of dicpip by agents in the tissue extracts other than succinic acid?

E3.4 Chloroplast activity in photosynthesis

The biologist investigating cell metabolism was likened earlier to a man trying to discover the workings of an engine that stops as soon as the casing is removed. In such a situation he can at least make some parts imitate their proper motion; he can push the pistons up and down and turn wheels to observe their functional relationship. The 'mechanism' of photosynthesis is far more complex than that of a motor car engine and much still remains obscure, but progress has been made by isolating parts of the system and making them work in an artificial yet informative situation.

Whereas respiration is essentially a process of oxidation, photosynthesis is one of reduction. Carbon dioxide is reduced, by a long series of reactions, to carbohydrates, fats, and other complex molecules. The green pigments of plants, the chlorophylls, have long been recognized as necessary parts in the mechanisms, but when these are extracted from leaves in dissolved form they are incapable of reducing anything. Examination of leaf cells (see Chapter 7) reveals the pigment confined to small corpuscle-like bodies or chloroplasts. Between pure pigments and intact plants, isolated chloroplasts are possible intermediate 'working parts' which may provide useful information.

In the following procedure a method for extracting intact chloroplasts is described so that you can investigate their reducing ability in the presence and absence of light. Complete photosynthesis in cell fragments is hardly to be expected and carbon dioxide cannot be used as a substrate. As in the previous investigation, dichlorophenolindophenol (dicpip) is useful because it can be seen to be reduced. Suspensions of chloroplasts must be free from mitochondria or the oxidation of succinic acid may cause the decolorization of dicpip. But we can distinguish photosynthetic activity from other metabolic activities by the effect of light. Chloroplasts, like mitochrondria, are delicate structures that need protection from osmotic damage and require an environment similar to that of the cell contents from which they have been extracted. A method of extraction is described; the design of possible experiments, using chloroplast suspensions, is suggested by the questions which follow. These should be read before starting to carry out the extraction process.

Procedure

1   Prepare phosphate buffer solution with added sucrose as described in the *Laboratory Book*, and place it in a refrigerator at 0° C.

2   Take 100 cm³ of this solution and dissolve 0·1 g of 2·6 dichlorophenolindophenol (dicpip) in it; store in a refrigerator.

3   Place the liquidizer attachment of an electric grinder/liquidizer in a refrigerator and, when cold, pour in 200 cm³ of buffer/sucrose solution.

4   Put a number of centrifuge tubes and a beaker in the refrigerator to cool.

5   Take a whole fresh lettuce (or an equivalent amount of leaves from another plant), cut out and discard the midribs, and let the remaining pieces of leaf fall into the solution in the liquidizer.

6   Run the liquidizer for 30 seconds only and rapidly pour the liquid into four layers of muslin (or nylon stocking) draped across the mouth of the cold beaker. Gather up the corners of the cloth; twist and squeeze the green suspension into the beaker. Discard the solid residue.

7   Stir the suspension quickly then pour it into cold centrifuge tubes. Fill to within 1 cm of the brim of each tube.

8   Spin the suspension in a centrifuge at 2,500 g for 5 minutes.

9   Pour away the green supernatant liquid. Add cold buffer/sucrose solution to the sedimented material and shake well to resuspend it.

10   Pour a sample of this suspension into a colorimeter sample tube and add dicpip solution. Insert a red filter and obtain a colorimeter reading at about the mid-point of the scale. A decrease in the blue colour of dicpip indicates its reduction.

Questions

a   Design and carry out an experiment to compare the effects of different light intensities on a chloroplast suspension containing dicpip as an indicator of chemical reduction. Record and discuss the results obtained.

b   What other factor or factors, associated with light, are likely to affect such an experiment? How would you eliminate such factors so that light intensity is the only one which differs?

c   How would you find out if the green suspension in fact contains chloroplasts?

d   Describe experiments you could perform to see if enzymes are involved in the reduction of dicpip by chloroplasts.

Bibliography

Baldwin, E. (1962) *The nature of biochemistry*. Cambridge University Press. (The oxidation of succinic acid in the Krebs cycle.)

Fogg, G. E. (1963) *The growth of plants*. Penguin. (The importance of the Hill reaction and the light reaction of photosynthesis.)

James, W. O. (1963) *An introduction to plant physiology*. 6th edition. Oxford University Press. (Respiratory rates and respiratory quotients.)

Synopsis

1 In preceding chapters, metabolic processes have been considered largely in isolation. But within living organisms these processes occur simultaneously. The effects of photosynthesis on the composition of the environment counter those of respiration; the net outcome can be seen in a study of compensation periods.

2 In natural communities animals and plants live together and their metabolic processes may have a complementary or counteracting effect on the environment. The situation can be investigated, using changes of carbon dioxide concentration in isolated, model communities.

3 Communities of plants and animals can survive without the addition of raw materials or removal of metabolites. This emphasizes the importance of element cycles in nature.

Chapter 8

Metabolism and the environment

Summary of practical work

section   topic

8.1   Compensation period

8.2   The interaction of plants and animals

8.1 Compensation period

Previous chapters have been concerned with the various ways in which individual organisms function. They have not only studied the organisms themselves individually, but have also investigated their internal processes such as breathing and digestion separately. The breakdown of complex problems into smaller, simpler ones is an essential preliminary to any scientific enquiry; it is part of scientific method. In a general biological study, however, it has one drawback in that it may encourage the idea that organs and organisms function independently of each other. For example, respiration and photosynthesis have been studied in different chapters. We know that all organisms respire and plants photosynthesize, and therefore we may presume that plants do both. The two processes can have an opposed or complementary effect on the environment as is shown in figure 84. Therefore when studying either process in a plant, we should always take into account the effect of the other.

This is all the more important when investigating plants in a natural, as opposed to a laboratory, situation. When light was required for the investigations of Chapter 7 it was provided continuously from an artificial source, but in natural conditions plants live in alternating periods of darkness and light, night and day. There is nothing wrong with the use of artificial, laboratory conditions so long as experimental results are not misinterpreted and regarded as natural events.

Suppose that the information given in figure 84 were gained by a series of separate laboratory experiments. It might be concluded that as A and B appear equal and opposite there is no net uptake or output from plants. It might further be pointed out that B occurs only in the light whereas A is unaffected by it, and in natural surroundings a plant is illuminated for only half its lifetime. As the situation becomes natural so it becomes more complex. The effects of two distinct metabolic processes must be considered in conditions of alternating darkness and light.

Techniques and materials, already described in previous chapters, may be useful in devising simple ways of investigating the compensating effects of respiration and photosynthesis.

Procedure

(*Note:* The questions following stage (7) should be read before starting practical work.)

1   Select two containers, large enough to contain several plants. The containers should be transparent, to allow the entry of light, but easy to seal, to prevent the entry or escape of gases. Figure 85 illustrates one example of such a container.

2   Because plants may differ widely in their ability to photosynthesize and respire, obtain collections of at least two different species of plant and set up at least two containers, each containing equal amounts of plants of the same kind.

3   Bubble air through a solution of bicarbonate/indicator until

Figure 84
A comparison of the effects of respiration and photosynthesis.

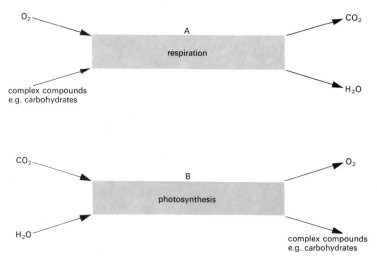

this is red. Devise a means of exposing samples of this indicator at various places inside the containers. One convenient method is shown in figure 85. Visking (cellulose) tubing will hold liquid yet allow the penetration of surrounding gases. The solution inside will not respond to changes in external gas composition as quickly as an indicator in open dishes, but these would have to be on the floor of the container whereas the Visking tubes can be suspended at any height. Seal the containers, which must be airtight.

4 Set aside a sample of bicarbonate/indicator in a sealed glass container so that it may be used later for comparing colour.

5 Ideally, the plants in their containers should be placed in the dark for about twelve hours. This is best done by putting them in a light-proof cupboard fitted with a lamp and time-switch. The latter should be set to switch the light *off* twelve hours before the next laboratory session. If a time-switch is not available it is best to cover the containers with an opaque box or cloth at the end of a school day. This will give a longer dark period than desirable, e.g. 5 p.m. to 9 a.m., but will nonetheless give useful results.

6 After the dark period note the yellow colour of the indicator and illuminate the containers equally. Record the power of the light source and measure the distance between it and the containers so that the experiment can be repeated exactly, if necessary.

7 Record the times taken for the yellow bicarbonate/indicator to return to its original red colour, in both containers. These times are known as *compensation periods*, and the stage at which carbon dioxide output is balanced by uptake is the *compensation point*.

Questions

a By what means can you ensure that the quantities of plants in both containers are equal?

b Is it better to stand the plants in soil or water? Give your reasons.

c Should steps be taken to maintain constant temperature in the containers and if so, why?

d The colour change of bicarbonate/indicator is gradual. It is difficult to assess the time it takes to become a certain shade of red again, accurately. What steps can you take to make the measurements more precise?

e Do the two or more species of plants used have the same compensation period? If there is a difference is it greater or less than the limits of accuracy of the timing method?

f Suppose, regardless of the answer to question (e), that some plant species differ in their compensation periods. Do you think that this would be important in a natural situation?

g Is the size of the container, in relation to the number of plants used, important? Would you expect different results using large containers with a few plants?

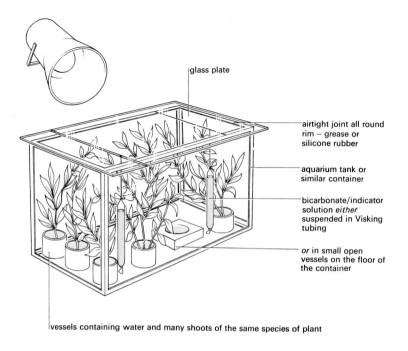

glass plate

airtight joint all round rim – grease or silicone rubber

aquarium tank or similar container

bicarbonate/indicator solution *either* suspended in Visking tubing

or in small open vessels on the floor of the container

vessels containing water and many shoots of the same species of plant

Two methods of supporting bicarbonate/indicator

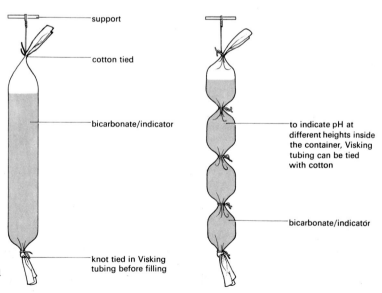

support

cotton tied

bicarbonate/indicator

knot tied in Visking tubing before filling

to indicate pH at different heights inside the container, Visking tubing can be tied with cotton

bicarbonate/indicator

Figure 85
An example of apparatus used to find the compensation period of plants.

h   How could you modify this experimental method to suit aquatic plants?

8.2 The interaction of plants and animals

The containers used in the previous section are not good models of natural communities. They contain plants of one species only and no animals, and they are completely enclosed. The effect caused by introducing animals is easy to foresee. Their respiratory activity will raise the carbon dioxide content of the environment and increase the compensation period of the community as a whole. What cannot be so easily foreseen is their quantitative influence. What proportion of animals to plants will make the compensation period infinitely long?

We can return, at this point, to the model communities set up as aquaria for the work of Chapter 1. If these have survived for some months without artificial maintenance, some sort of balance must have been struck between the plant and animal populations. The aquarium tanks are open to the air; would they survive if closed and sealed? It is easy to seal the tanks, but populations may take some time to change and it would be difficult to assess them in terms of numbers or biomass. If, instead, we use one or two organisms in small, sealed containers, we can gain information about a photosynthesis/respiratory balance. (Aquatic plants and animals can live in bicarbonate/indicator solutions.)

Procedure

1   Obtain small animals such as woodlice and so arrange them in a container that their total weight is between five and ten per cent of the fresh weight of the plants.
2   With the animals and plants together, repeat stages (2) to (7) of the previous section. Obtain a relative compensation period for the whole community.
3   Set up microaquaria (see figure 86). You can use large, corked test-tubes containing bicarbonate/indicator alone, pond weed, aquatic animals, and both pond weed and animals.
4   Place the tubes at equal distances from a light source and leave them for 24 hours. You may need a shield to avoid overheating.
5   Compare the colours of the indicator in the four microaquaria. Note whether the colour of the fourth is nearer that of the indicator containing animals only, or plants only.
6   Using this observation (5) set up a further series of microaquaria containing plants and animals in proportions which produce no change in the colour of the indicator during 24 hours of illumination.
7   Remove the contents of the most 'balanced' microaquaria and weigh the animals and plants.

Questions

a   What is the proportion of animals to plants used in stages (1) and (2), as determined by fresh weight?

b   Was the indicator restored to its red colour when the container was illuminated? From this observation state whether the community contained too many animals for continued survival or whether it could accept more.

c   What proportions of animals and plants, by weight, form a balanced microaquarium community in terms of carbon dioxide concentration?

d   Suppose that you wanted to set up a balanced aquarium community, would stages (3) to (7) above form a useful method for determining the proportion of animals and plants to be used? State the limitations of this technique and suggest improvements.

e   From these investigations, can you propose hypotheses concerning the proportions of animals and plants which form large, natural, balanced communities?

Figure 86
Microaquaria.

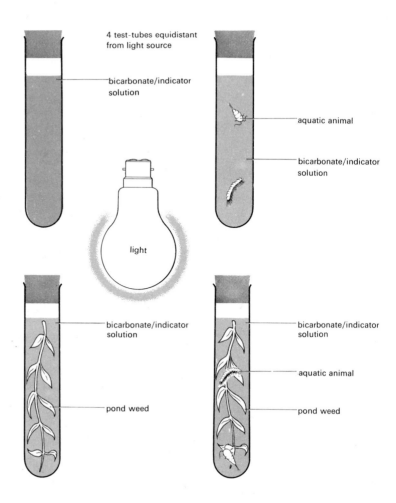

Isolated communities and chemical cycles

The investigations you have carried out in this chapter involve organisms in sealed glass containers. If properly balanced, such communities could survive indefinitely. We may well ask if such artificially controlled communities bear any resemblance to natural ones. When an aquarium or an indoor garden is sealed in a bottle, no material can enter or leave the enclosed environment, though light can penetrate the glass. Although this may seem unnatural such an arrangement serves as a model biosphere. The world is not sealed like a bottle, but it is more or less isolated and little material ever enters or leaves it. The layer or water, soil, and air covering the earth is extremely thin compared with the diameter. In figure 87 its layer is shown with a thickness equivalent to 130–160 km. Nearly all life is contained in a much shallower film only 1·5–3 km thick. From time to time meteorites fall on to the earth from outer space and some components of the biosphere such as hydrogen have atoms small enough to escape the earth's gravitational attraction. But these changes are trivial; the total amounts of all the elements on or near the surface of the world remain as constant as if they were sealed in some gigantic bottle.

The ceaseless cycle of life, germination, growth, reproduction, and, eventually, death and decay, involve a continual recombination of atoms. Combination and decomposition succeed each other over and over again. It may seem remarkable that organisms have not yet succeeded in synthesizing compounds incapable of decomposition. But if this were to occur on a large scale the atoms involved would be unavailable to subsequent life and the world would become poorer and poorer in these elements until eventually all life ceased. Happily, atoms continue to oscillate between the organic molecules of living organisms and the inorganic molecules of soil, sea, and atmosphere. In accomplishing changes they frequently go through repetitive series of combinations,

Figure 87
The Earth and its hydrosphere

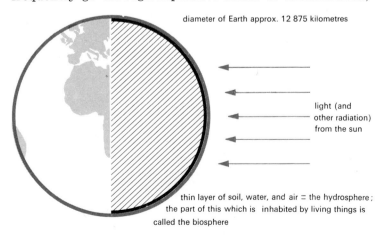

diameter of Earth approx. 12 875 kilometres

light (and other radiation) from the sun

thin layer of soil, water, and air = the hydrosphere; the part of this which is inhabited by living things is called the biosphere

hence it is customary to speak of atom or element *cycles*, the best known being those of carbon and nitrogen.

To follow atoms practically through a complete cycle of inorganic and organic compounds would be a lengthy and complex process. Rather than prescribe an investigation beyond the scope of available apparatus, the following questions call for the design of possible methods of examining various element cycles. Useful information may be gained from previous chapters as well as from textbooks.

Questions

a A method of tracing the entry of radioactive carbon atoms from the atmosphere into plant leaves has been described in Chapter 7. If you were provided with a sealed container, plants, animals, soil, and a source of radioactive carbon, describe how you would attempt to demonstrate a carbon cycle.

b Describe how, with apparatus and materials similar to those listed in (a), but using radioactive phosphorous in a solution as phosphate, you would investigate a phosphorus cycle. Radiophosphorus (^{32}P) has a short half-life of about 14 days; it emits more radioactivity, atom for atom, than ^{14}C. How would you modify your method accordingly?

c Refer to textbooks for details of the nitrogen cycle. There are no sources of radioactive nitrogen available. What techniques would you employ to investigate the nitrogen cycle?

d Of the many elements found in living creatures, five are often set apart as playing a prominent role, namely carbon, hydrogen, oxygen, nitrogen, and phosphorus. Describe hypothetical oxygen and hydrogen cycles. Is there any reason why the two should appear similar?

e 'The outer layers [of the earth] in their melted form, must have contained in addition to silicates much water and carbonate in solution, the whole being originally surrounded by an atmosphere of hydrogen and hydrides, CH_4, NH_3, H_2S and H_2O, though in far less quantities than now. This original atmosphere must have been modified in two particular ways. The actual crystallization of the crust must have forced into the atmosphere vast quantities of water vapour and carbon dioxide . . .' (from *The physical basis of life*, 1951, by J. D. Bernal, F.R.S.).
Supposing this to be an accurate description of the earth's original atmosphere, propose hypotheses to account for the change to its present constitution.

Bibliography

Baron, W. M. M. (1967) *Organization in plants*. Edward Arnold. (Provides an introduction to work on compensation periods.)
Bernal, J. D. (1951) *The physical basis of life*. Routledge & Kegan Paul. (Though somewhat dated, this is a concise statement of problems and hypotheses of the origin of life.)

Galston, A. W. (1961) *The life of the green plant*. Prentice-Hall. (Contains a brief introduction to the nitrogen cycle and mineral nutrition of plants.)
Stoneman, C. F. (1970) Nuffield Advanced Biological Science Topic Review *Metabolism*. Penguin.
Stoneman, C. F. (1970) Nuffield Advanced Biological Science Topic Review *Photosynthesis*. Penguin.

INDEX